ENVIRONMENTAL
AESTHETICS

Why have our public environments become ugly wastelands or banal blandscapes? How important is an aesthetically pleasing public environment to personal well-being?

Environmental Aesthetics explores the contributions made by a wide variety of disciplines to the conceptualization, conservation and design of environmental beauty in cities, rural areas and wilderness in the Western world. The book traces the history of aesthetic thought and practice, examining basic aesthetic concepts and the resultant implementation of aesthetic policy in the landscape. Discussing the psychology of human–environment relations and the influences of literary, legal and artistic activism, the author concludes with an analysis of the essential roles of public policy and planning.

This is the first comprehensive account of the new interdiscipline of environmental aesthetics. Unique in scope, the book dovetails concepts, methods and practice from disciplines as varied as architecture, art history, biology, environmental studies, forestry, geography, landscape design, law, literature, philosophy, psychology and urban planning. Equally at home with landscape art, psychological experiments, policy making and planning, the author brings us a step closer to understanding how our experience of city and country life can and should be improved.

J. Douglas Porteous is Professor of Geography, University of Victoria, Canada.

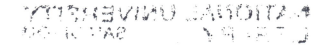

Other books by J. Douglas Porteous

The Company Town of Goole: An Essay in Urban Genesis (1969)
Canal Ports: The Urban Achievement of the Canal Age (1977)
Environment and Behavior: Planning and Everyday Urban Life (1977)
The Modernization of Easter Island (1981)
The Mells (1988)
Degrees of Freedom (1988)
Planned to Death: The Annihilation of a Place Called Howdendyke (1989)
Landscapes of the Mind: Worlds of Sense and Metaphor (1990)
Mindscapes (in Japanese, 1992)

ENVIRONMENTAL AESTHETICS

ideas, politics and planning

J. DOUGLAS PORTEOUS

LONDON AND NEW YORK

First published 1996
by Routledge
11 New Fetter Lane, London EC4P 4EE

Simultaneously published in the USA and Canada
by Routledge
29 West 35th Street, New York, NY 10001
Routledge is an International Thomson Publishing company

Typeset in Garamond by
Solidus (Bristol) Limited
Printed and bound in Great Britain by
Biddles Ltd, Guildford and King's Lynn

British Library Cataloguing in Publication Data
A catalogue record for this book is available from the British
Library

Library of Congress Cataloguing in Publication Data
Porteous, J. Douglas (John Douglas)
Environmental aesthetics: ideas, politics and planning / J.
Douglas Porteous.
p.   cm.
Includes bibliographical references and index.
1. Environment (Aesthetics)   2. Landscape architecture.
I. Title.
BH301.E56P67 1996   95-47890
111'.85—dc20

ISBN 0-415-13769-1
ISBN 0-415-13770-5 (pbk)

for my students

CONTENTS

List of plates, figures and tables   viii
Acknowledgements   xi
Preface   xv

1 INTRODUCTION   1
1.1 Environmental aesthetics   5
1.2 A short history of aesthetics   19
1.3 The origins of the aesthetic impulse   24
1.4 The senses   31

2 HUMANISTS   43
2.1 The humanist approach   46
2.2 A history of landscape taste   51
2.3 Landscapes   75
2.4 Some national contrasts   101

3 EXPERIMENTALISTS   111
3.1 The experimentalist approach   114
3.2 Properties of environments and persons   118
3.3 Scenery is good for you   132
3.4 Methodological problems   139

4 ACTIVISTS   149
4.1 Literary and design activism   152
4.2 Citizen action   162
4.3 Legal issues   169
4.4 Public policy   176

5 PLANNERS   189
5.1 Aesthetic landscape planning   193
5.2 Urban aesthetic planning   215
5.3 Environmental education   241
5.4 Aesthetic critique   249

AFTERWORD: Beyond aesthetics   261

Bibliography   266
Index   284

PLATES, FIGURES AND TABLES

PLATES

Fountains Abbey with water scene   xiv
Coastal forest, BC   2
Convolvulus   44
Easter Island *moai*   112
Logging truck   150
Public housing, Hong Kong   190
Bank of China building, Hong Kong   260

FIGURES

1.1 From standard of living, through quality of environment, to quality of life   7
1.2 A hierarchy of needs   8
1.3 Intangible relationships with environment   9
1.4 Structuring environmental aesthetics   14
1.5 Normal landscape colours   34
1.6 Colour dominants   35
1.7 Stepping-stones   38
1.8 Reduction of rich sensory experience   39
1.9 Visual aesthetics only   40
2.1 Landscape as background   53
2.2 Medieval and Renaissance gardens   54
2.3 Landscape by Claude Lorrain   56
2.4 Landscape by Nicolas Poussin   57
2.5 Landscape by Salvator Rosa   58
2.6 Landscape by Aert van der Neer   59
2.7 The pastoral   61
2.8 English rural landscape   62
2.9 The gothick   64
2.10 The pleasure of ruins   65
2.11 The picturesque   67
2.12 The bucolic   69

2.13 The wild   70
2.14 Aesthetico-political Romanticism   71
2.15 The surreal   74
2.16 American landscapes in the nineteenth century   80
2.17 Modern industry invades the countryside   80
2.18 The Baroque landscape   83
2.19 The uninterrupted gaze   84
2.20 The transformation of gardens and park at Stowe   85
2.21 Heveningham Hall, landscaped by Capability Brown in 1781   86
2.22 Classical pavilion in Fountains Abbey gardens   87
2.23 Landscape transformation: informality to irregularity   89
2.24 City of spires becomes city of factory chimneys   91
2.25 Medieval and Baroque city plans   93
2.26 Georgian elegance   94
2.27 Victorian exuberance   95
2.28 Modernist streetscape   97
2.29 Questioning the modernist aesthetic   98
2.30 Standardized post-modern architecture   99
2.31 English landscape tastes: the tamed   102
2.32 English landscape tastes: the historic   102
2.33 American landscape tastes: the vast   104
2.34 British Columbian landscape tastes   108
3.1 A schema of aesthetic response to environment   119
3.2 Characteristic pattern of aesthetic response   119
3.3 Coherence, complexity and mystery   122
3.4 Mystery yields to a complex, coherent view   123
3.5 A nested hierarchy of landscape preferences   124
3.6 A familiarity and preference matrix   126
3.7 Four primary specialist groups   129
3.8 Natural versus urban scenes   133
3.9 The nature tranquillity hypothesis   136
3.10 Sacred waterscape   137
3.11 Alternative approaches to research   144
3.12 Experiential research in the Arctic   145
4.1 A modern nowhereland   155
4.2 Blandscape   156
4.3 Anywhere, UK   157
4.4 Anywhere in the world   157
4.5 Canadian version of 'America the ugly'   158
4.6 Canadian version of 'God's own junkyard'   159
4.7 London's spires and domes obscured   161
4.8 Environmental protest   165
4.9 Wall art   166
4.10 Scenic conservation in Britain   179
4.11 A preservationist problem   181
4.12 Reflections on preservation   183

4.13 Solitary heritage building in a modern streetscape   184
4.14 Restoration of a streetscape   185
4.15 Site preservation problems   186
4.16 Forest clearcutting in British Columbia   187
5.1 The scenic resources of Scotland   197
5.2 Landscape values of the Coventry–Solihull–Warwickshire region   199
5.3 Preference evaluation in the Wye Valley   203
5.4 A model for visual quality management   204
5.5 Digital terrain model of projected visual impacts of alternative forestry practices   206
5.6 Creating a landscape restoration plan for Blenheim Park   210
5.7 Scenic resource analysis of two creeks, Niagara   213
5.8 Scenic assessment of the Lancaster Sound region, Canadian Arctic   214
5.9 The image of 'cityness'   216
5.10 Streetscape with terminal punctuation   222
5.11 Entrypoint   223
5.12 Enclosure   224
5.13 Openness to closedness   225
5.14 Modernist urban landscape   226
5.15 Post-modern quaintspace   227
5.16 The model San Francisco in the Berkeley Environmental Simulation Laboratory   228
5.17 The 'New-Yorkization' of San Francisco   229
5.18 A 'monster house'   238
5.19 Veneerism   238
5.20 Façadism   239
5.21 A landscape of power   255
5.22 Scenery of the future?   259

TABLES

1.1 A comparison of schemata   15
1.2 A typology of scientists   17
1.3 Left- and right-brain functions   28
1.4 Autocentric and allocentric senses   31
2.1 The influence of Renaissance art on later periods   60
3.1 The Kaplans' preference framework   121
3.2 Phillips' and Semple's environmental personality clusters   130
5.1 Comparison of US visual assessment and impact systems   211
5.2 Frequency of urban aesthetic control issues   236

ACKNOWLEDGEMENTS

First, I'd like to thank all those who provide me with the loving, supportive milieux so necessary for thinking, research, and writing: Carol and Gavin Porteous in Victoria, British Columbia; Priscilla Ewbank and Jon and Jesse Guy on Saturna Island, and Gwen Mell and the whole extended family in Yorkshire. It's no accident that many of the photographs in this volume derive from these gorgeous places.

Next, all those kind souls who, sometimes imperiously, have thrust books and articles at me with a 'read this': Paul Chamberlain, Pat Duffy, Harry Foster, Barbara Hodgins, Ratko Karolic, Rita Marks, Carol Porteous, Sandra Smith, and Linda Wilson, as well as all those who pop interesting items into my mailbox, most notably John Newcomb. Generations of students in my environmental aesthetics and other classes have also brought me ideas in many forms and provoked me to think again by asking awkward questions. I have benefited from the technical expertise of Susan Bannerman, Mike Edgell, Chuen-yan Lai, Sakyah Miller, Jennifer Pollard, Tanji Zumpano and the ever-helpful librarians at the University of Victoria Reference desk. Editors Tristan Palmer and Matthew Smith at Routledge, and their three anonymous readers, provided valuable critique and encouragement. And then there's that 'invisible college' of environmental aestheticians, people I rarely meet but am glad to know are there: Jay Appleton, the Kaplans, David Lowenthal, Ted Relph, Yi-Fu Tuan, Roger Ulrich and Erv Zube.

Places and music continue to support my creativity. Boat Pass (Saturna Island) and the Holy Mountain, Athos, renew my spirit. The sweet voices of Silvio Rodriguez and Victor Jara, alternating with Boccherini, keep my pen moving, though Max Bruch stops it. I couldn't get through the awful task of compiling the bibliography without John Lennon, The Doors and Bob Marley. And when my eyes are tired, Tennyson and Gerard Manley Hopkins lubricate them.

I thank especially Judy Simpson for her interest, tolerance and amazing ability to produce beautifully word processed copy from a scratchy handwritten manuscript not only accurately but as quick as boiled asparagus. Jill Jahansoozi, Marjie Lesko, Bev Ranniger, Ruth Steinfatt and Joan Gillie cheerfully helped finalize the typing.

The task of organizing and processing the 100 illustrations was carried out with speed, accuracy, dedication and flair by Ken Josephson. Ken also took four photographs (Figures 2.31, 4.9, 5.15, 5.18) and drew nineteen illustrations (Figures 1.1–1.4, 2.17, 2.20, 2.26, 3.1, 3.2, 3.5–3.9, 3.11, 4.7, 4.10, 5.1, 5.4). Ole Heggen drew Figures 1.7 and 3.3 and contributed the cartoon (Figure 5.22) in his inimitable style.

Finally, I thank those who generously read early versions of the manuscript: Martin Monkman, who read Chapters 1–4; and Phil Dearden and Larry McCann, who read appropriate sections of Chapter 5. Sometimes I have taken their advice.

Thank you all. The usual disclaimers apply. Direct all comments to: Doug Porteous, Saturna Island, BC, Canada V0N 2Y0.

CREDITS

Single illustration credits are found in the appropriate figure captions. Multiple credits follow:

Figures 5.2, 5.3 are reproduced courtesy *The Planner*, Journal of the Royal Town Planning Institute.

Figures 2.29, 4.5, 4.6, 4.16, 5.14 are reproduced courtesy the Department of Geography, University of Victoria, British Columbia.

Figures 1.5, 1.6, 1.7, 3.3, 4.1, 5.6, 5.7, 5.8, 5.9 and Table 5.1 are reproduced courtesy *The Western Geographical Series*, University of Victoria.

Figures 1.8, 1.9, 2.8, 2.9, 2.10, 2.11, 2.16, 2.18, 2.23, 2.28, 2.32, 2.33, 2.34, 3.4, 3.12, 4.2, 4.3, 4.4, 4.11, 4.12, 4.13, 4.15, 5.10, 5.11, 5.12, 5.13, 5.19, 5.20 are by the author.

Identifications and credits for chapter-opening photographs are as follows. Preface: Fountains Abbey, Yorkshire (author); Chapter 1: British Columbia coastal forest (Department of Geography, University of Victoria); Chapter 2: Convolvulus (author); Chapter 3: Easter Island *moai* (author). Chapter 4: Canadian logging truck (Department of Geography, University of Victoria); Chapter 5: Public Housing, Hong Kong (Hong Kong Government photo); Afterword: Bank of China Building, Hong Kong (author).

Quotations from the poetry of Gerard Manley Hopkins are taken from *Poems of Gerard Manley Hopkins* (Third edition, 1948) Oxford University Press.

Lines from *Slough* and *The Planster's Vision* are from *John Betjeman's Collected Poems* (Fourth edition, 1980) and are quoted by kind permission of John Murray (Publishers) Ltd.

Any errors or omissions in the above list or in the caption credits are entirely unintentional. Every effort has been made to seek permission to reproduce copyright material. If any proper acknowledgment has not been made, we would invite copyright holders to inform us of the oversight.

Fountains Abbey with water scene

PREFACE

Beauty is truth, truth beauty: that is all
Ye know on earth, and all ye need to know.
John Keats

I

Keats exaggerates. So does Proust when he tells us that the only paradises are those we have lost. Yet from such ideas we may generate myths to live by, which is the essence of both religion and humanism.

The world may indeed be lost, but it is not well lost. We may be enjoying the last act of a piece of tragic theatre. Aristotle explains that a tragedy begins with an overweening, aggressive act on the part of some hero, which angers the gods or disturbs the balance of nature. The restoration of order after such hubris is called nemesis, and involves catastrophe for the hero. The assault on nature, and thus on our own nature, by our urban-industrial culture may be just such a tragedy, and there is already overwhelming evidence that if the world ends, it will not be with nuclear war (Eliot's bang) but with the long-drawn-out whimper of ecocatastrophe.

But this seems too biblical a narrative, moving inexorably from creation to apocalypse. Surely something can be done to arrest this progress? Long ago, we would have awaited a *deus ex machina*, a god descending from the heavens to save us. Even postmodern theology, however, agrees with radical humanism that there remains only the existentialist answer: we have to save ourselves from an impending world-destruction which we ourselves have set in train.

How can this be done? Clearly, by coming to our senses and by first rejecting the absurd myths that the twentieth-century industrial world has lived by and tried to foist on everyone else: the primacy of economics; the inevitability of progress defined as more; the value of action over reflection; the importance of materialism; and the patently ridiculous belief that happiness derives from having rather than from being.

These traits are deeply embedded in the dominant cultures of the northern hemisphere. They will be rooted out not by fiat, by laws promulgated from top down for the better regulation of society, but by the autonomous actions of millions of individuals who experience personal revelation, make a conscious turn, and deliberately change their lives from materialism to spirituality, from passiveness to creativity, from having to being.

II

What has all this to do with environmental aesthetics? First, neither Capitalist nor Communist myth-makers ever devoted much serious attention to what the world would be like after their particular ideology had triumphed. Like the heaven of some biblical religion, no doubt all would be sweetness and light. Meanwhile, it was paradise postponed as each ideology grappled with its contradictions and declared that the duty of the masses was not to reason why but to continue the struggle toward a future era of peace and plenitude. In much the same way we shortchange our children by forcing upon them a curriculum crammed with short-term, 'useful' subjects, while neglecting the fact that a sense of personal worth often derives not from what society has learned to call 'success', but from the satisfactions of relationships and of creativity in art, music, and literature. When all the world's problems are on their way to being solved (and, let's face it, many politicians, business people, and scientists devoutly fear such a consummation) art, literature, and music will be all we have left with which to occupy ourselves.

So why not, as part of our personal turn toward creativity and being, engage with aesthetics now? And as the environment is clearly in danger unless we change our lives of endless having, the development of a better appreciation of environmental aesthetics is one step toward the restoration of world-order and earth-order. Technological lunacies from Chernobyl to *Exxon Valdez* have caused the public to question the assurances of their politico-economic leaders that profits and 'progress' matter more than environmental integrity. Such integrity has a strong aesthetic component; man-made environmental disasters are both dangerous and ugly. And when ugly begins to mean dangerous, there is an increased likelihood that the quest for beauty may form part of the spearhead of the quest to restore a sound ecological world-order.

III

If aesthetics is so important, and apparently so inextricably linked with ethics, why has the subject, nay even the very word 'aesthetics', been so lamentably neglected in twentieth-century industrial culture? In the Marxist world, aesthetics clearly had to serve ideology directly. The Western world, led first by Victorian Britain and then by an unreflecting, triumphalist United States, readily took to heart Thomas Jefferson's dictum that we should embrace the practical arts before the so-called decorative ones. This is profoundly false advice based on a foolish distinction which, in earlier times and in many existing Third World cultures, simply cannot be made. Industrial society is nothing if not fragmented; holistic visions are difficult for us.

Aesthetics, therefore, apparently concerned only with surface appearances, has clearly been deemed to be of little importance in comparison with the business of making a living which, as we are beginning to realize, so often means making a killing. Thus aesthetics was relegated to a very minor position in the urban-industrial pantheon. It could contribute little to material 'progress', a myth which Northrop Frye calls a secular parody of the biblical promises of life hereafter.

Consequently, aesthetic issues have been important only to a tiny minority of the population, whether artists, intellectuals, or the very rich whose contribution often seems restricted to the turning of works of art into mere commodities. (For the masses, natural or historic beauties have long been commercialized.) Aesthetics, then, was accused of elitism and was associated with capital-c Culture, or even Kultur, which in the societies we choose to call democracies has been under a cloud for over half a century. Aesthetics, as a word in common usage, has been neglected even longer, having come into undeserved disrepute in the 1890s with the public humiliation of the 'aesthete' Oscar Wilde for homosexual practices.

Since that time an interest in aesthetics has been firmly associated with the epicene, effeminate, and effete, especially in men. We still live in a patriarchal society, and *macho* values prevail. Men must not seem to be sensitive; they can safely cultivate an eye for flowers only when they retire and thus become socially and symbolically unsexed. Even the word beauty, along with its partners, culture and aesthetics, went out of fashion except in highly political situations such as 'beauty contests', whose contestants until recently had to be healthy, slim, young, white females. And in the 1960s began a veritable cult of ugliness, symbolized perhaps by a close-up of the incomparable face of Mick Jagger sounding its barbaric yawp across the world.

IV

Nowhere is the cult of ugliness so obvious as in the United States, a most disquieting notion for those who are not wholly enamoured of that country's claim to represent civilization, be the custodian of Western values, and possess the right to make over the world-order in its own image:

> Here is something that the psychologists have so far neglected: the love of ugliness for its own sake, the lust to make the world intolerable. Its habitat is the United States. Out of the melting pot emerges a race that hates beauty as it hates truth.

Thus H. L. Mencken in 1927, and a host of environmental commentators since.

Yet all this is now over. Even the American public now suspects military-industrial-political talk of progress. We suspect science and technology of being inhumane. We suspect that a better world might well be made when ethics, aesthetics, and survival are linked together. Ugly is no longer cool.

Sensitivity among males is now countenanced; William James's tender-minded now share the stage with the tough-minded, and may usurp them. Feminism uproots centuries-old myths and promotes the earth, once again, as mother. State capitalism has collapsed; there is even hope that Western capitalism might be humanized. The two former world-systems are at last beginning to realize that they have in common an urgent need to cooperate in the preservation and promotion of an ecologically-sound earth-system where the ugly and dangerous are eliminated in favour of the beauty of a harmoniously-functioning earth supporting the creativity of individuals.

v

At the end of the twentieth century, *mirabile dictu*, we have restored to ourselves the right to have vision. As more individuals come to discard their old values, their endless progress towards more, an ethic of environmental humility will join with an aesthetics of environmental harmony as a central focus in our lives. This book is written in the context of such a vision. In the words of the psalmist:

> Earth might be fair
> And all men glad and wise.

1

INTRODUCTION

Coastal forest, BC

Glory be to God for dappled things –
   For skies of couple-colour as a brinded cow;
      For rose-moles all in stipple upon trout that swim;
Fresh fire-coal chestnut-falls; finches' wings;
   Landscape plotted and pieced . . .
                    Gerard Manley Hopkins

The initial goal of this book is to further the development of a new interdiscipline, Environmental Aesthetics, which is as yet in its infancy. To that end, I provide an overview of a wide array of very disparate literatures, from architecture to forestry and tourism, from art through biology to urban planning, all of which contribute conceptually, theoretically or empirically to the nascent interdiscipline. An overview, however, is not enough. Accordingly, I also develop below a typological framework which clusters the many relevant disciplines, emphasizes linkages between them, and relates work in the humanities and sciences to public activities and professional planning. Furthermore, this theoretical framework acts as a kind of hypothesis, for it suggests the necessity for a free flow of information and ideas between the Humanists, Experimentalists, Activists and Planners who make up the four chief approaches to Environmental Aesthetics, and asks what impediments lie in the way of this intercommunication. Unimpeded interaction between the four groups is, I believe, a prerequisite for my final goal, which is to make some small contribution towards the creation of a more aesthetically pleasing world.

At this early stage the reader should note that the material I discuss is largely derived from literature in English, although some French and German work has also been used. Consequently, the book will concentrate almost wholly on Western approaches to Environmental Aesthetics.

To grasp the nature and context of Environmental Aesthetics at the turn of the twentieth century, it is first necessary to understand that the aesthetics of environment is generally judged to be of far less importance in the public domain than other issues such as standard of living; yet personal aesthetics is of great moment to the individual (Chapter 1.1). This initial section continues with a comparative survey of several recent suggestions on how best to organize the new interdiscipline. The structure chosen is characterized by a reasonable balance between the qualitative and the quantitative, the theoretical and the applied, and provides the organizational plan for the rest of the book.

The chapter then examines three contextual issues. A short history of the development of aesthetics as a discipline from medieval to modern times (1.2) includes a discussion of basic aesthetic concepts. This is followed by a discussion of two major theories of the origins of the aesthetic impulse (1.3). Finally, a necessary survey of the major sensory modalities (vision, sound, smell, tactility) through which we monitor environment is provided (1.4). All this introductory material sets the stage for the four interrelated chapters (2–5) which make up the bulk of this book.

1.1 ENVIRONMENTAL AESTHETICS

Considered in terms of private satisfactions, aesthetics is clearly of vital importance to the human sense of well-being. Industries catering to aesthetic satisfactions derivable from dwellings, pets, cars and body grooming, for example, are thriving economic enterprises.

North Americans in particular are willing to spend a considerable proportion of their disposable incomes on body, hair, and clothing fashions and grooming. The jogging craze and the growing rash of fitness centres exemplify a 'body beautiful' aesthetic as well as an ethic of preventive medicine. Tanning and tattooing parlours as well as 'colour draping' and 'aesthetics studios' exemplify a growing awareness of body image in an age in which cultural literacy appears to be in decline (Hirsch 1987). And with the development of fashion, tattooing and cosmetics, the body becomes a landscape to be adorned and enhanced. Paradoxically, there is concurrently a strong concern for scrubbing the body clean, and cleanliness is still next to what today passes for godliness. In our drive to eliminate bodily odours, in particular, we betray considerable ignorance of history (in which almost everyone stank), of the aesthetic sensibilities of other cultures, in which body smells are often important referents, and of the sensory pleasures to be obtained from these intimate body odours.

Why this frenzy of activity? In large part because our culture has developed an exaggerated respect for packaging, for shadow rather than substance, and is willing to reward its citizens for their personal appearance alone; this is not new, for Aristotle made the same observation. As Ackerman (1990: 271) reports:

> The sad truth is that attractive people do better in school, where they receive more help, better grades, and less punishment; at work, where they are rewarded with higher pay, more prestigious jobs, and faster promotions; in finding mates, where they tend to be in control of the relationship and make most of the decisions; and among total strangers, who assume them to be interesting, honest, virtuous, and successful. After all, in fairy tales, the first stories most of us hear, the heroes are handsome, the heroines are beautiful, and the wicked sots are ugly. Children learn implicitly that good people are beautiful and bad people are ugly, and society restates that message in many subtle ways as they grow older.

In particular, advertising creates ideal body-images which the majority of the population cannot attain, but somehow feel they should. They then fall prey to reduced self-esteem and into the traps of the sharks and vultures who purvey self-help books, diet programmes, cosmetic surgery, slenderizing systems, and a host of other lost causes which together make up a multibillion dollar industry based wholly on a false and historically extremely recent body aesthetic.

Small, ugly, bald, lame presidents of the United States were not wholly absent before John F. Kennedy and the mass television culture. To elect one now, or a Rubenesque female perhaps, would be difficult. It is significant that whereas women, visible minorities, and the handicapped have made great strides in self-emancipation and therefore in self-esteem, protest movements by the 'fat' and the 'ugly' have been limited and mostly unsuccessful. Female, black, or deaf are obviously physical characteristics, difficult to alter; fat and ugly are value judgements, like 'weed'. Social reality is an agreed-upon fiction which works to the benefit of an aesthetically-attractive minority.

Our houses, too, are like our bodies the shrines of aesthetic concern and care. All houses could well be painted black or white externally. Gardens and lawns could be ignored or eliminated, to the chagrin of numerous popular magazines of the 'homes and gardens' variety. Interior decoration and visual display do not necessarily have to be prominent social concerns. Only in a culture which has fetishized scenic views will hilltop sites and the upper floors of apartment towers fetch the highest prices. (Before the elevator, of course, it was the urban poor who had to trudge upstairs to tiny garrets while the rich enjoyed all the advantages of spacious apartments at ground level.)

In our dwellings we are wont to keep pets. One of the reasons for this activity is the aesthetic satisfactions gained thereby. Beyond the visual, cats are often valued for their tactility, dogs for kinaesthetic sensations. Human babies may be treated in this way ('isn't she lovely . . .?'), as are cars.

The public context

When we move from the private to the public domain, however, too often we find that aesthetic concerns quickly drop to a level of marginal interest. 'In the play of forces that govern the world' says Tuan (1994), 'aesthetic . . . ideas rarely have a major role.' Private affluence amid public squalor is a fact of life in the US and Britain and if, as Pawley (1974) suggests, we face an increasingly 'private future', the entrenchment of this attitude in actual landscape seems inevitable. Even in terms of current environmental problems, aesthetic issues are of low priority, and are frequently considered a mere luxury in comparison with pressing ecological concerns. This is especially so in the US, where the overwhelming thrust has been to obey the Jeffersonian imperative to embrace first the 'practical arts', and only later the 'decorative' ones. In Lynch's (1976) words, 'Esthetics is often considered a kind of froth, difficult to analyze, easy to blow away.' One crass US senator was wont to speak of 'ass-thetics.'

Why this neglect? Figure 1.1 suggests that increasingly affluent Westerners pursue a personal, family, or intergenerational progression from overwhelming concern with 'standard of living' through later concern for 'quality of environment' and, finally, 'quality of life'. This progression has a number of salient characteristics. First, it is a movement from 'having' (the accumulation of tangible things, often paradoxically resulting in an impoverished life involving large doses of low-grade electronic pleasure) to 'being', where the quantitative assembling of owned objects gives way to the more sophisti-

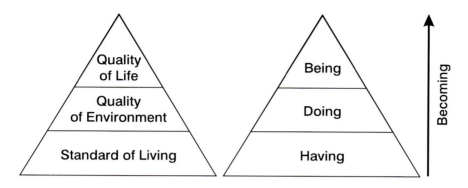

Figure 1.1 From standard of living, through quality of environment, to quality of life

cated, qualitative, enjoyment of aesthetic intangibles (artwork, cleanish air, wilderness experiences, the pleasures of creativity).

Second, it is clear that there is a strong class component in this progression. Aristocratic and bourgeois elites, taking standard of living and quality of environment for granted, have long concerned themselves with quality of life. With the general embourgeoisement of Western populations in the twentieth century, elitist aesthetic and lifestyle patterns have trickled down the social hierarchy to be emulated, in a more modest fashion, by an increasingly larger segment of the population. Thus what were once eighteenth-century aristocratic tastes for smooth lawns and adorned private carriages have become the commoner's Sunday morning burdens of cutting the lawn and washing the car.

At the turn of the millennium it is probably safe to say that the masses of the Western world are emerging from concerns with standard of living to confront both global and private issues of environmental quality. In the twenty-first century, perhaps, with ecological, economic, and military problems on their way to being solved, we may well be able to concentrate upon quality of life while at the same time assisting Third World peoples to achieve decent standards of living in environments of acceptable quality.

A third feature of the concept is its broad similarity to the well-known hierarchy of needs propounded over a generation ago by the humanistic psychologist Maslow (1954). Figure 1.2 suggests a framework of human needs which form an ascending hierarchy from the immediate and basic to the ultimate and sophisticated. Clearly, we must first satisfy physiological needs, such as those for food, water, and shelter, before dealing with safety needs, which include protection from physical harm, the maintenance of privacy, and the opportunity to reduce psychic threats from others.

Beyond this first pair of bodily needs lies a second pair of social needs. Affiliation needs include the need for group membership, and thus involve friendship, love, and family. Given these, we are more easily able to pursue our esteem needs, which relate to personal integrity via both self-evaluation and the perceived esteem of others for oneself.

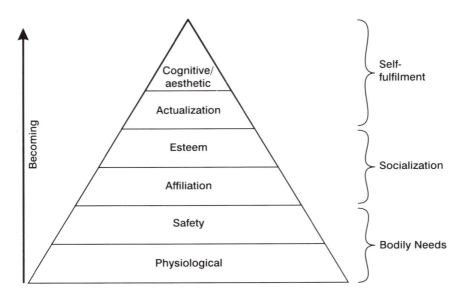

Figure 1.2 A hierarchy of needs (after Maslow 1954)

Towards the top of the hierarchy we encounter actualization needs, or the need for self-fulfilment according to one's capacities. And on achieving the very peak of the pyramid, all other needs satisfied, we may be able to indulge our cognitive and aesthetic needs, which involve our enjoyment of learning and our personal concepts of beauty.

Maslow's expressed goal was to facilitate the movement of both individual and society up the hierarchy toward the peaks of creative fulfilment. An analysis of the 'self-help' genre of books, avidly purchased by the middle classes of North America, indeed suggests that the lower two levels of the hierarchy, which we may label 'consumerism' in this context, are not problematic for the majority. The bulk of self-help books focus upon the middle tier in their emphasis on family, social, sexual, and organizational relationships. By the 1990s, however, a considerable salient had been created in the third tier with the rapid growth in numbers of bestsellers dealing with a wide variety of aspects of creativity and spirituality.

It is at this level that we are able to situate aesthetics in the context of other critically-important intangibles (Porteous 1993). Figure 1.3 centres upon the notion of being-in-the-world, whose centrality depends upon the four somewhat Jungian supports of thought (mind), feeling (heart), intuition (soul) and sensation (the gates of the body). Sensation relates directly to aesthetics, traditionally concerned chiefly with surface appearance rather than meaning. Meaning, however, is paramount in the other areas. Intuition leads us to spirituality or metaphysics, which in environmental terms may be expressed in peak experiences, epiphanies, or 'oceanic' feelings generated by landscape experiences. Feeling may involve attachment (or its opposite), and thus sense of place and perhaps a deep love for particular places. Finally, thought leads us toward ethics, whereby we consider environments in terms of moral good.

To illustrate this schema I will take the example of southern Moresby Island in the Queen Charlotte archipelago off the northern British Columbia coast immediately south of Alaska. One's aesthetic sensibilities are immediately aroused by the island's lush but severe landscapes composed chiefly of forested mountains and wooded inlets. Aesthetically, here we have an opportunity to enjoy nature through all our senses. Spiritual intuition tells us that this is a sacred space (Graber 1976), a now-uninhabited morsel of almost pristine wilderness. Our ethical sense draws support from ecological, aesthetic, and heritage values in a drive to have the area designated as a national park (partially successful in the 1980s). Meanwhile, to the Haida nation, which once inhabited the now-lonely inlets where skewed totems and mossed longhouses moulder, the area is ancestrally sacred; the Haida are attached to this land by deeply-felt ties.

South Moresby, of course, was fought over by preservationists and logging companies, with several levels of government as participants and arbiters. Here we learn the valuable lesson that the aesthetics of landscape

Figure 1.3 Intangible relationships with environment

must not only be considered in the context of place, ethics, and spirituality (loosely, 'meaning') but is also inevitably embedded in a social, economic, and political matrix.

In terms of political economy, however, aesthetics still lags behind utilitarian, instrumental, and ecological concerns. Nevertheless, by the late 1980s the growth of the tourist industry had led British Columbian leaders, in common with politicians the world over, to reconsider landscapes as revenue-generators in their own right. The rights of logging and mining companies may therefore, in the future, be curtailed in favour of the maintenance of a landscape which retains the capacity to promote aesthetic pleasure. Such rethinking, however, is likely at first to be directed only to a few highly-visible areas. We are a long way from an ideal situation where aesthetics might be given equal consideration with economics or politics (Levy-Leboyer 1977).

In this connection, a statement by Prall (1929: 1) sardonically voiced nearly three generations ago, provides a fitting final commentary on my attempt to situate environmental aesthetics in its societal context at the turn of the twentieth century:

> When all the places in the world are ... immediately to be reached by light-wave transportation or perfected instruments of distance vision; when all the problems connected with having the fare to pay for these instantaneous journeys are completely solved; when there are no more wars or national rivalries or passports or customs regulations to complicate the overcoming of natural barriers by the raising of artificial ones; when men have full knowledge of their bodily mechanisms and their health ... and when ... they shall have found out how to live together in society – when all this is accomplished, if we are not to sit down and wish for the return of our practical difficulties to give us occupation or for death to relieve our boredom, aesthetic matters will have to be taken seriously.

Structuring Environmental Aesthetics

Having situated Environmental Aesthetics sociopolitically, I move on to disciplinary structuration. The search for order is a vital human concern. Order in the field of environmental aesthetics is especially necessary because of the rampant growth of empirical research in the area since the 1960s, a growth poised uncertainly atop an almost total theoretical vacuum. Moreover, the increasingly wide scope of both research and practice, and lack of sufficient bonds between the two, demand some attempt to provide at least a workable framework for the field.

Few significant attempts had been made prior to the 1980s. Chalmers (1978) proposed a pioneering research typology which made tentative but unclear distinctions between basic and applied research and the use of quantitative and non-quantitative methods. Quantitative approaches included both psychological work on aesthetic response and the environmental determinants of this response (Chalmers' 'aesthetic appreciation' category) and 'aesthetic preference' studies in which respondent preference is the yardstick, but again environmental variables, rather than personal ones, are the most tested. Non-quantitative methods included work on 'innate attachment' and 'history, literature, and art', which involved the mining of secondary sources. Chalmers himself questions his distinction between aesthetic appreciation and preference, and generally the attempt was seen as pioneering but unsatisfactory.

Other partial and unsatisfactory attempts to structure environmental aesthetics appeared in the early 1990s. Tuan (1993) does not really provide a structure; as usual, his work is idiosyncratic. Berleant (1992) is similarly wide-ranging, philosophical, and understructured. Bourassa's (1991) work is structured too simply into biological, cultural, and personal units (see also Bailly, Raffestan and Raymond 1980). Nasar (1988), similarly tripartite, has sections logically arranged in terms of theory, research, and applications. None of these provides a structure which is comprehensive and coherent; fortunately, several more useful structuration attempts appeared in the interval.

In the early 1980s I noted that 'a synthesis of current empirical work in environmental aesthetics is long overdue' (Porteous 1982: 53). And in an astounding example of synchronicity, no fewer than three independent proposals for the categorization of the new interdiscipline appeared in 1982. Punter (1982), taking a philosophical approach, identified three major paradigms and suggested a materialist perspective. Zube, Sell and Taylor (1982), taking a theoretical approach based on rigorous quantitative analysis of the available literature, proposed four paradigms and created a model of 'landscape perception'. Taking a more problem-oriented approach, I also proposed four paradigms and outlined their interaction in terms of theory generation and application.

Punter's 'Landscape Aesthetics'

While acknowledging that the area is difficult to categorize, Punter states that 'three interdisciplinary perspectives or paradigms are immediately apparent' (p. 102).

The first, landscape perception, focuses upon the mechanics of how we perceive landscape and upon the links between perception, comprehension, preference and action. Although a great deal of work has been done on the perception, cognition, evaluation, and assessment of environmental qualities, this research cannot stand alone because it tends to ignore the meaning of the landscape for the observer. Meaning is paramount in the second paradigm, landscape appreciation, which deals, rather qualitatively, with landscape as an expression of cultures, lifestyles, and values. The third paradigm, known as the landscape or visual quality approach, is an explicitly aesthetic approach which deals with formal sensory-aesthetic qualities, particularly in town-scape, and has direct application in urban planning.

Perhaps because of this rather urbanist bias, Punter attempts to integrate these three threads via a materialist perspective, a kind of neo-Marxist aesthetic. He is particularly severe on both critics and academic humanists for their 'privileged indifference' and detachment. Very properly, he wishes to see informed landscape appreciation as an everyday experience for everyone, rather than chiefly for specially-educated critics. And in a very useful countering of traditionalist aesthetic orientations he proposes that a materialist aesthetics would probe beneath aesthetic surfaces to uncover the moral substrata and ethical implications lurking beneath surficial patinas. In this exhortation not to pity the plumage and forget the dying bird, Punter's work complements that of Relph (1981) on the aesthetics and ethics of the North American fast food landscape.

Zube, Sell and Taylor's 'Landscape Perception'

Zube and his colleagues were interested in discovering what kind of approaches had been used in 'assessing perceived landscape values'. Twenty American, British and Canadian periodicals were reviewed for the period 1965–80, and 160 relevant articles found. Only articles which included a definite method for assessing landscape quality were included, a criterion which may have excluded a number of humanistic papers which displayed no explicit methods or assessment techniques. Four major paradigms emerged from this process.

At one extreme, the expert paradigm involves the evaluation of landscape quality by skilled, trained observers. Skills are derived from practical training in art, design, or resource management, and aesthetic values are implicit rather than explicit. This approach implies that 'wise resource management techniques may be assumed to have intrinsic [positive] aesthetic effects'. At the opposite extreme lies the experiential approach, which considers land-scape values to be based on experience of human-landscape interaction, where both are continually shaped and reshaped by the interactive process. This paradigm concentrates heavily on sociocultural, historic, and individual experience. Unlike the expert approach, where the expert is a detached observer with purely utilitarian goals, here the observer becomes a subjective, active, participant.

The role of observer is dominant again in two rather more psychologically-oriented paradigms, the psychophysical and the cognitive. Like the expert approach, the psychophysical is much concerned with the evaluation of landscape quality for planning purposes. Here, however, expertise is eschewed in favour of batteries of tests which monitor the landscape evaluations of the general public. Here the observer becomes respondent, and external landscape properties are assumed to have a correlational or even stimulus-response relationship with observer evaluations and behaviour. The cognitive psychological approach, in contrast, is more oriented toward basic research. Often an experimental approach, it is much concerned with the sociocultural properties of the observer, who is seen as an information processor. The search for the human meaning assigned to landscapes or landscape properties is a major concern.

Zube *et al.* also discuss some of the similarities, differences and linkages between the four paradigms. For example, the expert and psychophysical approaches stress applicability, while the expert and cognitive approaches have similar views on the importance of certain compositional characteristics of landscape, such as edges, mystery, and complexity. A major interest in the meaning of landscapes and of human-landscape interactions is common to both the cognitive and experiential approaches. As the authors note, there are many other areas of similarity and connection.

Finally, in what appears to be an extension of Zube *et al.*'s work, Daniel and Vining (1983) propose a very similar methodological typology, varying mainly in the dichotomization of Zube *et al.*'s expert paradigm into formal (landscape architecture) and ecological (biological) approaches.

Porteous's 'Environmental Aesthetics'

Unlike the Zube *et al.* approach, which is based on a rigorous impersonal analysis of a sample of the available literature, my own schema (Porteous 1982) emerged from the combination of extensive personal reading and intuition. Published in the *Journal of Environmental Psychology* in 1982, its logic derives not simply from the literature itself, but from the goals that modern scientists publicly espouse.

Two major philosophical goals have emerged in the social sciences during the last generation. One is the quest for rigour, whereby many social scientists have attempted to emulate as closely as possible their colleagues in the physical sciences (the 'physics envy' model). Quantitative, behavioural approaches are indicated. The second is the quest for relevance, pursued with varying rigour by those concerned with applicability and policy-making. Rigour regardless of relevance was once pursued with vigour in universities. The 1970s and 1980s saw the 'rush to relevance' on the part of many academics. In the 1990s, standing a little apart from both fads, we may hope to achieve the goal of an earlier, prescient observer, 'relevance with as much rigour as possible' (Roepke 1977).

Using the two criteria of relevance and rigour as axes on a graph, four distinct but interconnected approaches to the investigation of environmental aesthetics may be discerned. Relevance, referring to the immediacy of the approach to the solving of current environmental problems, is clearly of great importance in view of the widespread belief that the aesthetic quality of environment is undergoing considerable decline through the multidimensional assaults of urban-industrial societies. Rigour, which refers to scientific

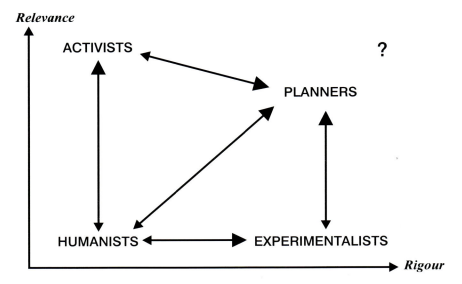

Figure 1.4 Structuring environmental aesthetics

theory-building and testing, is also regarded as vital by the growing number of environmental designers, managers, and policy-makers who look to social science for a conceptual and informational background to action.

Figure 1.4 situates the four paradigms both with respect to these two criteria and in relation to each other. They range from a humanist or purist, contemplative, approach which emphasizes the surficially unique while at the same time seeking universals intuitively, thus necessarily eschewing immediate relevance and scientific positivism, to the extremes of relevance characterized by environmental activists ('act now!') and the extreme rigour of experimentalist social scientists ('before we can change the world we must first understand it'). Environmental planners (managers, designers, policy-makers, etc.), who confront immediate issues and now usually have a fairly rigorous scientific or social scientific training, occupy an intermediate, central position. No group, as yet, has reached the extreme position of 'high relevance with extreme rigour'. Although this may be a consummation devoutly to be wished, in practice it is likely to be difficult to attain.

It should be noted that the approaches are located on the diagram in rather extreme, and hence isolated, positions. As a corrective, it should be understood that these approaches are tendencies only; they are not hermetic compartments. Individuals working in the field of environmental aesthetics frequently stray across their very permeable boundaries, and these inter-disciplinary trespassers are often those most able to illuminate the common problems that confront all groups. In other words, while the designation of four paradigms has a definite basis in the reality of the literature, it may well be regarded as a heuristic device for the reader's better understanding of what should be a holistic interdiscipline. (We are all, to some extent, in academic ruts, but we should never forget that a rut is a grave with the ends kicked out.)

Comparison of schemata

As Table 1.1 suggests, there are extensive areas of agreement among the Punter, Zube and Porteous schemata. Although each schema was developed quite independently, using very different methods, the Zube *et al.* and Porteous frameworks bear a strong similarity. In contrast, other observers, whose work has not been directly concerned with creating a framework for environmental aesthetics, have tended to reiterate Chalmers' threefold division into humanistic ('qualitative'), scientific ('quantitative') and applied categories.

The chief problems with most of these categorizations are that they are static, apolitical and more concerned with the academic literature than with the attitudes of those who write it. They are static because, except for Zube *et al.*, they pay little attention to interactions between paradigms. They are apolitical because, except for Punter, they fail to acknowledge that environmental aesthetics, as with any other aspect of modern life, has an important sociopolitical dimension. My own framework attempts to overcome these difficulties.

First, it postulates necessary and essential interconnections between the four paradigms. It assumes that no academic works in total isolation from those whose philosophical and methodological approaches markedly differ. It also suggests that there should be strong connections between academics (humanists and experimentalists) and both professionals (planners) and the public (activists). In other words, the schema has normative overtones.

The four approaches, in fact, far from being seen as antagonistic, competing, hermetic compartments, should be regarded as complementary. I postulate a continual flow and reflux of information and concepts between the paradigms. Ideally, the humanist group could be regarded as an important

Table 1.1 A comparison of schemata

Porteous (1982)	Punter (1982)	Zube *et al.* (1982)	Chalmers (1978)
Humanists	landscape interpretation	experiential	history, literature and art/innate attachment
Experimentalists	landscape perception	cognitive/ psychophysical	aesthetic appreciation/aesthetic preference
Activists	–	–	–
Planners	landscape quality	experts/psycho- physical	expertise/aesthetic preference

generator of basic concepts which experimentalists test and refine and which activists use to further their specific causes. All three might be expected to impact both directly and via each other on the planners, who evaluate these inputs before recommending or undertaking action in the public domain.

Second, my schema acknowledges the importance of the political. Indeed, it suggests that political aspects of environmental aesthetics are so important that they deserve their own place in the framework (activists) alongside the humanist, scientific, and applied paradigms. If aesthetic considerations are ever to be implemented in a concrete way in city, countryside and wilderness, the planners who perform this work must operate within a framework of legislation which in turn is often the end-product of lobbying and protest on the part of environmental activists.

Third, the schema is based not merely on the academic literature in environmental aesthetics, but on the attitudes and goals (rigour, relevance) of scientists, professionals, and the public as represented by activists. In this connection the literature on how scientists think about and practise their research is worth a brief excursus.

Humans are arrant compartmentalizers. There are two types of people: those who divide everything into two and those who don't. Academia is rampant with this kind of thinking: Arts and Science; Natural Science and Social Science; Pure and Applied, and the like. This world-view often includes the tendency to believe that science is monolithic, that there is such a thing as 'the scientific method', and that hard, quantitative science is the acme of academic endeavour.

Nothing could be further from the truth. Enough studies and personal confessions have emerged by now to show that the tenor of an individual's work is strongly influenced by upbringing, socialization, personality, and contingency. Numerous attempts have been made to produce typologies of scientists themselves, rather than of the literature they produce (although the one, of course, depends on the other).

Hudson (1966), for example, distinguishes between convergers (who have a penchant for the impersonal analysis of facts) and divergers (who are holistic generalists), while Maslow (1966) speaks of healthy, creative scientists as opposed to unhealthy, compulsive scientists, who are characterized by neurotic levels of precision, control, quantification, and proof. Other typologies (Mitroff and Kilmann 1978) suggest three or even four types of scientist, although even here the important contrast is between creative, speculative, integrative pattern-finders and 'technicians' who can neither recognize problems easily nor solve them without prompting. The difference is aptly illustrated in a passage from Canadian novelist Robertson Davies (1980: 182), where a diverger rebukes a converger:

> You like the mind to be a neat machine, equipped to work efficiently, if narrowly, and with no extra bits or useless parts. I like the mind to be a dustbin of scraps of brilliant fabric, odd gems, worthless but fascinating curiosities, tinsel, quaint bits of carving, and a reasonable

amount of healthy dirt. Shake the machine and it goes out of order; shake the dustbin and it adjusts itself beautifully to its new position.

One of the most interesting four-level typologies is that of Mitroff and Kilmann (1978). It derives from the work of Jung, who postulated a basic framework in which Thinking is opposed to Feeling, and Sensing to Intuition. According to Jung, individuals have a preference for taking in data via either Sensing (sense data) or Intuition (akin to imagination). Similarly, an individual may process information and make decisions via either Thinking (impersonal, formal and theoretical) or Feeling (using personal value judgements). Using this Jungian framework, Mitroff and Kilmann have created a fourfold typology of scientists: the Analytical Scientist (Sensing and Thinking); the Conceptual Theorist (Intuition and Thinking); the Conceptual Humanist (Intuition and Feeling); and the Particular Humanist (Sensing and Feeling).

Salient characteristics of the research approaches of these Mitrovian scientist-types are suggested in Table 1.2. The methods and goals of persons who typify these approaches to knowledge will clearly differ markedly. Currently, the Analytical Scientist occupies a privileged position, product of a major public relations effort over the last several centuries, which emphasizes the importance of precise, expert, exact, impersonal knowledge based on careful, controlled observation or experimentation. Both Conceptual Theorists and Conceptual Humanists are open to a much broader array of methods, the first with the goal of expanding and improving the world's stock of conceptual schemata, the second with the goal of fostering human growth and development. Particular Humanists may reject normal scientific methods entirely in favour of literature, art, music or action as ways of knowing; their goal is to help individuals (or, in planning, particular places) towards self-knowledge and self-determination.

Although my Rigour–Relevance typology of environmental aesthetics was developed before I encountered the Mitroff and Kilmann schema, the two clearly enrich each other. Looking back at Figure 1.4, my Humanists are likely to be Conceptual Theorists and/or Conceptual Humanists, imaginative, speculative generalists with a holistic bent, capable of generating

Table 1.2 A typology of scientists

Analytical Scientist	Conceptual Theorist	Conceptual Humanist	Particular Humanist
apolitical	apolitical	political	political
impersonal	impersonal	personal	personal
expert	generalist	generalist	localist
methodical	speculative	speculative	biased
logical	imaginative	imaginative	focused
specific	holistic	holistic	specific

Source: data derived from Mitroff and Kilmann (1978)

innovative concepts. These concepts may then be operationalized, tested and refined by the Experimentalists (Analytical Scientists). Once well known, the Humanists' global concepts may be localized and given a particularistic problem orientation by Activists (Particular Humanists), who are unlikely to be scientists at all.

Here my schema departs significantly from that of Mitroff and Kilmann, which was concerned wholly with the orientations of academic scientists. In contrast, the Rigour–Relevance schema involves also the public (or at least those protestors, environmentalists, lobbyists and legislators who may be termed Activists) and professional planners. The latter group, who aim for high levels of both rigour and relevance, fall into none of the Mitrovian categories. Rather, the good planner feeds on information from all the other types and must therefore have considerable familiarity with each, while yet not identifying wholly with any. While conceptualists generate ideas, experimentalists test them, and activists rephrase them as specific problems, the planner is asked to provide solutions and oversee implementation, a delicate balancing act. Many planners are indeed trained as Analytical Scientists, but contact with reality, often in the shape of political and business pressures, generally leads to their abandonment of this position.

On a personal note, this book has been conceived by a Conceptual Theorist (Porteous 1982) who has had some training as an Analytical Scientist and has actively participated as a Particular Humanist (Porteous 1989). Towards the end of the final chapter, however, and in the Afterword, the book becomes rather more Conceptual Humanist in tone.

It is my opinion that the Rigour–Relevance schema remains as yet the most useful framework for understanding environmental aesthetics. Accordingly, its four approaches (Humanist, Experimentalist, Activist and Planner) are used as the means of structuring Chapters 2 to 5 below. Before considering these approaches in detail, however, it is useful to outline the history of aesthetics, speculate on the origins of the aesthetic impulse, and discuss the nature of the senses.

1.2 A SHORT HISTORY OF AESTHETICS

The word 'aesthetics' derives from the Greek *aisthanesthai*, 'to perceive', and *aistheta*, 'things perceptible', as contrasted with things immaterial. Hence the *Oxford English Dictionary* is correct in defining aesthetics as 'knowledge derived from the senses'. A related definition, that of the philosopher Kant, regards aesthetics as 'the science of the conditions of sensuous perception'. This is too broad, for by the twentieth century the emphasis had moved from sense to sensibility with definitions such as 'the apparent embodiment of emotion in art', or the *New English Dictionary*'s 'philosophy or theory of taste, or the perception of the beautiful in nature and art'.

Indeed, the notion of beauty seems deeply embedded in our conception of aesthetics. The question 'what is beauty?' has been at the centre of aesthetic theory since Classical Greece. Plato, for example, ruled that form, rather than content, made a work of art beautiful, and asserted that beauty was independent of truth or usefulness. His pupil Aristotle was convinced that the three essential components of beauty were wholeness (*integras*), harmony (*consonantia*) and radiance (*claritas*). The notions of balance, harmony, proportion, and order, and the concepts of the Golden Mean and 'nothing in excess' also emerged from this cultural source.

Medieval aesthetics, in contrast, was a branch of theology. Classical theories of beauty and art gave way to a conception of beauty as the radiance of truth (*splendor veritatis*) shining through the mundane artistic or natural symbol, and reflecting God. With the Renaissance, however, aesthetics was resurrected as one of the normative disciplines, namely ethics, logic, and aesthetics, or goodness, truth, and beauty, later to be confused by Keats' 'Beauty is truth, truth beauty'.

Modern aesthetics can be traced back only to the eighteenth century, when the word was reinvented by the philosopher Gottlieb Baumgarten (1714–62). Etymologically, as we have seen, aesthetics originally was 'the study of perception', but Baumgarten's penchant for poetry and the arts in general led him to redefine the subject as 'the theory of the liberal arts ... the science of sensory cognition'. He took this line because of his belief that the perfection of sensory awareness could be found in the enhanced perception of beauty.

Contemporaneously, but without the use of Baumgarten's neologism, the aesthetics of sensory beauty were being worked out in eighteenth-century England. This is the great watershed in the history of aesthetics; much of our aesthetic theory derives from this period and place. Britain at this time was deluged with treatises on the 'laws' of beauty, which were derived from Classical models and the Renaissance revival. These works often took the form of technical manuals on how to achieve beauty in the various art forms.

Theory was equally rampant. The first systematic statements came from Addison and Burke. Burke (1909, originally 1757) was one of the first thinkers after the Scientific Revolution to downplay reason and argument in favour of examining 'the passions in our own breasts'. Addison (Green 1856) contrasted the pleasures of the imagination (refined) with the pleasures of the senses (gross) and the pleasures of understanding (cold). Both believed that 'almost everything about us' can be aesthetically valuable, including nature as well as art. Aesthetic distinctions, then, were to be made on the basis of felt experience.

Eighteenth-century aestheticians were also responsible for breaking down the preoccupation with beauty which had bedevilled aesthetics for the previous two thousand years. In this they were truly modern. Beauty, in fact, came to be associated with uniformity, harmony, variety, balance, and proportion, all the Classical aesthetic components. But it was obvious that things which were non-uniform, lacked variety, or seemed disproportionate (such as deserts or mountain ranges) still aroused the emotions. Hence a degree of non-beauty, or even downright ugliness, took its place in the aesthetic canon. Finally, the eighteenth century was of importance in its development of what is known as the 'aesthetic attitude', one of the chief components of which is the notion that the aesthetician is not motivated by advantage, but attends to the perceptual object 'for its own sake'.

The term 'aesthetics', however, was slow to be accepted except in a highly technical sense as 'the philosophy of taste'. Indeed, early in the nineteenth century English-speakers reviled the neologism 'aesthetics' as mere German metaphysics, and De Quincey (1839) satirized its corollary, 'good taste', in his famous essay *Murder considered as one of the the Fine Arts*. Despite taking on a somewhat pejorative sense during the 'naughty nineties', when aesthetes were pilloried in Gilbert and Sullivan's comic opera *Patience*, the word gradually took on its broader modern scope and was in common use by the early twentieth century. It is possible that the negative connotations of the word, associated invariably with Oscar Wilde (Gaunt 1988), led early twentieth-century philosophers to retreat from an aesthetics which confronted life in general to a far more narrow confrontation with art alone. By the mid-twentieth century certain aesthetic philosophers had come to consider that aesthetics dealt not with the theory of beauty but with the theory of art.

Beyond philosophy and the fine arts, aesthetics has been severely neglected in twentieth-century academe. Tuan (1989: 233) believes this neglect relates to our strange disregard of surface phenomena, even though such phenomena are one of our chief sources of pleasure: 'The scholar's neglect and suspicion of surface phenomena is a consequence of a dichotomy in Western thought between surface and depth, sensory appreciation and intellectual understanding, with bias against the first of the two terms.'

By the late twentieth century, even aesthetic philosophers had all but forgotten that aesthetics once dealt with nature as well as human artifacts. The basic questions they now ask, however, are of value whatever the object of aesthetic interest. It is sufficient at this point to outline some of the issues which aesthetic philosophers address (Saw and Osborne 1968: 27):

When you commit yourself to a judgement that a given object is beautiful, are you demanding the agreement of other people? ... Do you feel that you are saying *more than* 'I enjoyed it'? If so, what more is involved? In short, do you think that judgements made upon works of art are matters of taste and that *de gustibus non est disputandum*, or that considered opinion upon works of art may be supported by argument? If you hold the former view, do you hold the corollary that any opinion is as good as any other or do you admit experts in criticism? If so, how would you describe the expert? Is it contradictory to hold that some opinions are worth more than others?

And so on. These questions, although limited here to a consideration of works of art alone, are the key to understanding much of the debate which has enlivened the course of environmental aesthetics in the last two decades.

Some aesthetic concepts

It is not my intention to deal in depth with aesthetic philosophy. Some basic concepts are, however, worth noting at this point, as they are essential to understanding what follows.

All aesthetic questions involve *preference* (Prall 1929). Aesthetics also involves the art of *discrimination*, of making *judgements*. With the growing ability to distinguish good from bad, one develops *taste*. Tastes may be collective, or intersubjective, but their verification can only be subjective; we may be told that a sculpture is beautiful but we can only tell if it is indeed so by looking at it ourselves. Aesthetics is basic to human nature; artistic ability has a naturalness or even an inevitability (Prall 1929: 45). According to Prall, 'the so-called aesthetic motive is one of the outstanding springs of action among primitive men', resulting in their drive to beautify the useful, as, for example, in the working of sword hilts. Further, aesthetic activity focuses upon objects as they appear directly to the senses; aesthetics is not concerned with the origin or purpose of an object. To appreciate the object's *surface* (hue, pitch, rhythm, contour etc.) does not involve any understanding of its use. Finally, Prall notes that an appreciation of nature, rather than works of art, in this manner is relatively modern, but he dismisses the satisfactions of nature as being 'only accidentally and not intentionally beautiful' (p. 301).

Fifty years later, Beardsley's (1982) aesthetics contrasts sharply with Prall's. Beardsley begins his work with the argument between Consolidated Edison and conservationists, before a New York State commission, over the possible siting of a nuclear power station on the Hudson River. Unfortunately, the aesthetics of nature then all but disappears from the book.

What Beardsley does have to offer is a strong feeling for what might be termed 'aesthetic activism'. He tries to operationalize aesthetics for modern society by creating such notions as 'national aesthetic wealth' (the totality of all aesthetically valuable objects), 'aesthetic welfare' (all the aesthetic levels of experience of members of a given society at a given time), and 'aesthetic justice' (which deals with the fairer distribution of aesthetic welfare). These notions have immediately practical applications. Aesthetic welfare suggests that some effort should be made to redress the imbalance in modern cities between neighbourhoods of aesthetic affluence and those of aesthetic poverty. In superpower terms, I strongly question the American insistence, common in the 1980s, on the winnability of a 'limited nuclear war'. Such a war was to be limited to Europe. One can hardly take seriously the American establishment's willingness to destroy the wealth of aesthetic treasure lying between Edinburgh, Leningrad, Athens, and Lisbon in favour of saving, for a grateful posterity, the aesthetic splendours of Kansas City and Omaha. Nevertheless, it is clear that to give weight to such opinions, and to increase aesthetic appreciation in general, so that more harmonious environments may eventually be created, there must be an immense effort in aesthetic education. Given the consumerist and militarist stances of most major nations, such a redirection of effort hardly seems likely.

Finally, it seems useful at this point to note the distinctions often made between sensory, formal, and symbolic aesthetics. Sensory aesthetics is concerned with the pleasurableness of the sensations one receives from the environment; it is concerned with sounds, colours, textures, and smells. Formal aesthetics is more concerned with the appreciation of the shapes, rhythms, complexities and sequences of the visual world. Symbolic aesthetics involves the appreciation of the meanings of the environments that give people pleasure, or otherwise (Lang 1988). In a more condensed manner, Hospers (1946) speaks about the 'thin sense' (physical appearance, i.e. sensory and formal) and the 'thick sense' (expressive values, or meaning) of aesthetics. This book will cover all types of aesthetics, but it is clear that discussions of architecture or psychological experimentation might well concentrate on formal aesthetics, while humanistic or activist approaches will be far more concerned with meaning. Meanings are important, because they underly the drive for environmental planning, yet recent writers agree that symbolic aesthetics has too often been ignored outside art history (Nohl 1980, Lang 1988). Greenbie (1982), indeed, argues strongly that much aesthetic significance, especially of little-visited landscapes such as wilderness, is largely symbolic. Gertrude Stein was wrong; a rose is most definitely not merely a rose.

The aesthetics of environment

Very few modern aesthetic philosophers have cared to venture beyond the art gallery, unless to direct their gaze at individual architectural creations. While philosophers of aesthetics are happy to write essays on 'vulgarity' or 'the concept of the interesting', they tend to behave as if neither natural forms nor man-made large-scale environments (i.e. 'landscapes' or 'townscapes') have any aesthetic value.

Hepburn (1963, 1968) feels that the withdrawal of aesthetic philosophers from the outside world came with the post-Romantic, post-Wordsworthian feeling that nature, far from being our educator, is essentially indifferent and unmeaningful. Science, in the shape of microscope and telescope, has produced 'some bewilderment and loss of nerve over the aesthetic interpretation of nature'.

Philosophers are keen to emphasize the differences between natural objects and works of art. Works of art, for example, are 'framed' in some way, whereas natural objects are frameless. That which lies beyond the frame cannot be part of the aesthetic experience, whereas in a natural environment extraneous signals may intrude, as when aircraft noise penetrates a sylvan scene. Hepburn has clearly not considered the music of John Cage. Further, there appears to be no body of 'critical literature' or 'systematic description' to provide a 'background' for interpretation. Whereas art appreciation is promoted by built-in guidelines to interpretation, contextual controls for our response which are deliberately placed there by the creator, no such guides can exist in nature. A landscape is an 'unframed ordinary object', and rigorous modern aesthetics simply cannot handle objects which have no author, no system underlying their making, and which belong to no known school.

Hepburn's chief problem with environmental aesthetics, however, is the problem of detachment. One can step back from a painting. In contrast, in a landscape the viewer is involved, environed, enwrapped, surrounded (Collot 1986). He can go in, and is likely to experience not only the landscape but perhaps also himself in an unusual and vivid way. Yet far from being a handicap, this quality may be of extreme value, as we shall see when discussing humanist landscape appreciation.

Few philosophers have followed Hepburn's attempt to come to grips with environmental aesthetics. A decade later, Rose (1976) was commenting that 'the aesthetic properties of nature and the widespread appreciation of them are in large part ignored by present-day aesthetic theories'.

If the philosophy of aesthetics cannot help us understand our reactions to environment, we will clearly have to turn elsewhere.

1.3 THE ORIGINS OF THE AESTHETIC IMPULSE

Compared with earlier centuries, the twentieth century has shown an almost complete lack of interest in the underlying 'laws' or principles of aesthetics. This is particularly true of environmental aesthetics. This absence of interest has varied origins, including our knowledge of the failure of earlier attempts to discover aesthetic laws, the failure of a scientific age to understand the well-springs of creativity, an insistence on the part of certain aestheticians that 'beauty' is an entirely subjective experience, the growing precision of scientific method which has led to an increasing isolation of the arts, the growth of the financial nexus in the fine art world, and a general late twentieth-century desire to be free of restrictive doctrines.

On the other hand, a search for some degree of generalization becomes warranted when we realize that human beings continually search for order, and that even scientists are not immune to aesthetic judgements, as when a mathematician or experimentalist praises an 'elegant' solution or research design. Further, we have a great deal of evidence to suggest that aesthetic experience, although undoubtedly subjective as experienced, is also inter-subjective and can be generalized in a number of ways.

Eric Newton (1950), the English art critic, for example, notes that the perception of beauty varies according to a whole array of positional, temporal, and personality variables (p. 18):

> Beauty is a desirable commodity. But not all men are equally suscepti-ble to it. Nor are all men agreed about its abode. Moreover, it varies with period. It is subject to the laws that govern fashion . . . It also varies with its geographical position . . . Variations in national or racial standards of beauty are as noticeable as period standards.

Newton's general statements have been supported in numerous psycho-logical experiments and anthropological treatises which demonstrate that the appreciation of beauty varies according to a wide array of social variables.

Yet there is evidence for more universal agreement. Experimental aes-thetics, beginning with Fechner in the 1860s, has demonstrated substantial agreement among subjects who consistently matched mood adjectives with passages of music from Brahms (stately), Mendelssohn (sprightly), Mozart (wistful), Tchaikovsky (vigorous), and the like. Even when the more cacophonous and less familiar music of Hindemith, Bartok, Berg, and Stravinsky was used, the consensus labelling by mood adjectives was in the order of 70–97 per cent (Prall 1929). Similar matchings of adjectives with paintings have produced substantive agreement far above chance levels. To test whether this is merely a matter of culture, American students were confronted with such relatively little-known styles of music as Beduin, Javanese, Korean, and Maori, with the same results. Further, American, Chinese, and other subjects agreed in their assessments of a similar array of paintings. It is clear from these experiments that we must go beyond *de gustibus* and seek some underlying principles.

Two major attempts have been made to generate an underlying theory of the origin of the aesthetic impulse. One, the work of Appleton (1975a), a geographer, deals chiefly with nonurban landscapes and pre-civilized human-kind. The other, generated by Smith (1977), a psychologist, is more concerned with urban environments created in the last several thousand years.

Appleton's *Experience of Landscape*

Appleton's (1975a) theory rests on the notion that the roots of aesthetic appreciation lie in human biology. This, of course, is not a new idea (Kuhn 1968). Appleton's work is based on the great burgeoning of ethological research in the 1960s, a sociobiology which emphasized nature rather than nurture. The 1960s were rife with images of naked apes obeying territorial imperatives, and careful ethological observations in the field suggested that animals have a preoccupation with survival which manifests itself in nest-making, shelter-seeking, food-finding, and the desire to see without being seen.

Appleton takes this a step further by suggesting that humans have an 'atavistic sensitivity' to the landscape in terms of survival. Our reactions to landscape are partly inborn, and thus aesthetic feeling is at least partially based in biology. If the behaviour-mechanisms which govern our relation-ship to environment are inborn, they must clearly be activated and reinforced by environmental experience. Thus, 'if he is to experience landscape aesthetically, an observer must seek to recreate something of that primitive relationship which links a creature with its habitat' (p. viii). In trying to answer the question 'What do we like about a landscape and why do we like it?' Appleton posits a general 'habitat theory' and a more specific 'prospect and refuge theory'.

Habitat theory simply suggests that human beings experience pleasure in and satisfaction with landscapes insofar as these environments are perceived to be conducive to the realization of their biological needs. We appreciate most those environments which display the characteristics most favourable to our survival. Aesthetic satisfaction, then, is 'a spontaneous reaction to landscape as a habitat' (p. 70). In other words, it is a live-in, rather than a look-at, experience. Appleton suggests that even when we are able to control environments to such an extent that they become safe, this biological mechanism remains with us, genetically embedded, for use when required. We may thus view a landscape as satisfactory but be able to assign no immediate reason for our reaction, for problems of survival seem far off.

If habitat theory is generally 'about the ability of a place to satisfy all of our biological needs' (p. 70), prospect–refuge theory is based specifically on a hunting existence. Ethologists have noted that it is important for animals to see without being seen. Animals are quick to exploit the advantages latent in their surroundings, for to do so is a matter of life and death. The hunter needs to be able to view the prey (prospect) while hiding (refuge) until the final dash is made. The huntee must have wide vistas all around (prospect), plus the chance of getting away to a hidden place (refuge). In order to find such places creatures actively explore their environments seeking landscapes of prospect (the unimpeded opportunity to see) and refuge (the opportunity to hide).

Prospect and refuge are clearly complementary. The most satisfying landscapes are high in both, though the relative lack of one may be offset by strength in the other. The natural landscape type with one of the better balances between prospect and refuge is the African savannah, where *Homo sapiens* is thought to have originated. A number of writers have speculated that our origins in the savannah have led to an innate predisposition to respond positively to savannah-like environments (Orians 1980, Balling and Falk 1982, Ulrich 1983, Woodcock 1984, S. Kaplan 1987).

In one study (Balling and Falk 1982) children were found to prefer savannah to any other landscape type. Lyons (1983), however, explains this preference in terms of respondents' experiences of and familiarity with savannah landscapes. Yet later work by Orians (1986) has suggested more detailed mechanisms, in particular certain shapes and arrangements of foliage which could have served to indicate the suitability of an area for human habitation. Most recently, Orians and Heerwagen (1992) have found agreement among American, Argentine, and Australian respondents in their preferences for tree shapes; all subjects preferred examples of *Acacia tortilis*, a typical savannah species which indicates high quality savannah, rather than examples which came from lower quality savannahs.

From related evolutionary arguments the psychologists S. Kaplan and R. Kaplan (1982) have developed a theory of environmental preference which roots preference in the adaptive value afforded by particular settings (Gifford 1987). This comes close to Appleton's more intuitively-derived position.

The Kaplans, along with Ulrich (1983), have attempted to confirm their speculations concerning survival-based modern aesthetic values by empirical research (see Chapter 3.3, 3.4 below). Rather earlier, Appleton also attempted to forge a direct link between his ethological theory and modern aesthetics by analysing a wealth of literary and artistic material, as well as man-made landscape types. Paintings are found to be prospect-dominant, refuge-dominant, or balanced. Whole landscapes can be seen in these terms, as in landscape gardening and the deliberate creation of prospect (rolling country with wide views) and refuge (coverts and woods) in the fox-hunting landscapes of Midland England.

In paintings, as in landscapes, prospects are delineated as light, open, convex, and smooth, with an emphasis on distance and panorama. In contrast, refuge-dominant landscapes are concave, dark, treed, irregular, enclosed, and nearby. Woods, caves, rocks ('Rock of Ages, cleft for me'), ships and buildings become archetypal symbols of refuge, while crags, towers, and viewpoints generally are regarded as prospect symbols. Balanced landscapes, in these terms, are frequently found in the paintings of Constable and Gainsborough (Figures 2.12, 5.21).

The evolutionary or genetic approach (Appleton 1990) to aesthetics has been tested in several ways. Hildebrand (1991), for example, used Appleton's prospect-refuge theory to analyse the consistent appeal of houses designed by Frank Lloyd Wright. Heerwagen and Orians (1993) have tested Appleton's hypothesis by using surveys associated with photographic surrogates, by analysing the working notebooks of the eighteenth-century landscape architect Humphrey Repton, and by assessing landscape paintings. The latter study demonstrated gender differences; females were more strongly associated with refuge landscapes than were males. In a thorough appraisal of this research thrust, Ulrich (1993) concludes that biologically prepared learning plays a considerable role in the human response to unthreatening natural landscapes. In particular, genetic predispositions are important in liking and approach responses, in stress recovery responses (see Chapter 3 below) and in enhancing high-order cognitive functioning. The landscapes most preferred internationally are characterized by moderate to high depth or openness, relatively smooth or uniform-length grassy ground surfaces, and scattered trees or tree-clumps. Preference for such landscapes is much enhanced if they contain water. This is a description of an East African savannah. It also describes Chinese and medieval European hunting parks, eighteenth-century English landscape gardens, and modern urban parks (see Chapter 2 below).

Smith's *Syntax of Cities*

One sustained critique of Appleton's habitat theory is that it adapts less than well to the urban environment in which the bulk of the world's population now live. This problem is confronted by Smith whose goal in *The Syntax of Cities* (1977) was to go beyond the process of perception outlined in his earlier *Dynamics of Urbanism* (1974) by grappling with value systems and aesthetic judgements. His immediate objective was very much an urbanist's restatement of Appleton, for he wished 'to apply different aspects of psychology to the problem of urban design in an attempt to probe into how it is that some towns and cities offer pleasure in many dimensions' (p. 3). Smith also wished 'to provide convincing psychological reasons why urban design should be liberated from the straitjacket of contemporary planning and architectural philosophy' (p. 7).

His approach is via neuropsychology. Late twentieth-century neuropsychological theory suggests that the human forebrain can best be considered as a limbic system and a frontal neocortex. The limbic system is the seat of the emotions; it deals with the non-rational. In contrast, the neocortex is the thinking brain, but is itself divided into lateral hemispheres with rather different functions (Table 1.3). In short, the left hemisphere is Apollonian: verbal, mathematical, logical, deductive, and oriented towards the external environment ('outward bound'), whereas the right hemisphere is Dionysian: holistic, intuitive, spatial, pattern-recognizing, and concerned with inner spaces ('inward bound').

Table 1.3 Left- and right-brain functions

Left hemisphere	Right hemisphere
verbal language	eidetic (image) language
detail	pattern recognition
linear	geometric, three-dimensional
manipulative	reactive
orthodox	creative
behavioural (do)	experiential (feel)
rational (scientific)	emotional (artistic)
analytical (tree)	synthetical (forest)
departmentalizes	emphasizes relationships
form (spherical ball)	colour (blue ball)
concrete (a shoe is a shoe)	associative (a shoe: let's walk)
tangible (seeing is believing)	intuitive (that's possible)
time (next thing to do)	space (enjoy where you're at)

Smith argues that we have a physiological need to satisfy all three systems' demands for input, and that individual well-being depends upon the balance of input into the three areas. He uses a wealth of examples to demonstrate how a variety of urban landscape features creates a complex aesthetic potential which then provides psychological rewards via inputs into the three neurological processors.

Unfortunately, the Modernist (or Internationalist) movement in architecture, with its penchant for minimalist cubes devoid of decoration, has succeeded in dominating the cores of most of the world's larger cities (Figures 2.28, 5.14). According to Smith, this development of placeless identikit identitowns is 'left-cerebral dominant' and thus violently skews the modern city's potential for generating aesthetic satisfaction in its denizens.

In particular, modernist building styles deny the limbic brain, also known as the old, the visceral, or the gut brain. This system is less concerned with finer aesthetic evaluations, which can be left to the neocortex. Rather, it deals with emotional responses, biological needs (rather like Appleton's notions of safety and hunting), and carries primitive memories and archetypal values in the Jungian mode. Its need is for the exotic, bright colour, glitter, beat, rhythm, gigantism, massiveness, and repeated patterns. In short, the limbic system responds positively to what Smith calls 'pacer architecture' but what modernist architects would regard as vulgarity and 'glitz'. (Some of these needs have, not incidentally, been satisfied by the Post-modernist movement which was in its infancy when Smith was developing this theme in the early 1970s.)

A wide variety of supportive material is drawn upon. Sensory deprivation experiments (Porteous 1977) suggest that while we seek optimum levels of stimulation, we prefer overload to deprivation. Rapoport and Kantor's (1967) classic paper 'Complexity and ambiguity in environmental design' uses a vast array of psychological and other literature to assert the proposition that both humans and other animals prefer complexity and ambiguity in their everyday environments and that deprivation of these features can be detrimental to both physical and mental well-being. Developmental work on the physiology of monotony suggests that the initial environmental context of the neonate partly determines how the adult will see or otherwise sense its future environments; cats raised in wholly horizontal environments during the critical period of plasticity of the neocortical neurons grew up able to see only horizontal environmental features, for the neurons of the visual cortex had adapted to the nature of the visual input. The movement toward enriched sensory environments for human infants thus makes neurological, as well as common, sense. Finally, the early Post-modernist postures of Venturi and his associates (1972) amply support Smith's assertion that the limbic brain is starved in modernist city cores and suburbs and seeks outlet elsewhere.

To satisfy these physiological cravings, denied in modern identitowns, European tourists flock to the creative jumble of medieval Italian hilltowns while North Americans indulge in the limbic sensationalism of nocturnal Las Vegas. For Smith 'Las Vegas represents a concentrated eruption of limbic desires, unrestrained by the finer sensibilities of the higher brain' (p. 214). Critics of this tinseltown might well agree!

The crux of Smith's argument, then, is that aesthetic satisfaction derives from a dialectical interplay between the complexity and chaos desired by the limbic system and the linear order demanded by the left neocortical hemisphere, both being balanced and structured by the pattern-dominant right hemisphere. Modern cities, left-cerebral dominant, deny this balance and reduce aesthetic well-being, which their inhabitants must perforce seek elsewhere. And until this urban design imbalance is rectified, Smith urges us on to Las Vegas with cries of 'Viva vulgarity!'

Problems and critique

Both the Appleton and Smith theories purport to be universalist. In an age of narrow and cautious judgements such sweeping views are to be welcomed. Both theories, however, suffer from similar problems, notably the highly selective use of data, zoomorphist behavioural extrapolation from the non-human to the human, and ethnocentrism.

Habitat and prospect-and-refuge theories are heavily based on rural imagery, and Appleton's attempt to translate these concepts into urban terms is less than convincing. The universality of these biologically-based theories has been seriously questioned by those who tend towards the cultural side of the hoary nature–nurture controversy (Morgan 1978). Appleton, moreover, is highly selective of data. Though ranging through architecture, painting, literature and landscape design, many of his examples are chosen from the late eighteenth-century Picturesque period when the prevailing aesthetic in most of these modes does happen to fit the theory. Morgan (1978) suggests that this period is exceptional rather than typical and thus not a fit basis for universalist theory-building. Technically, Morgan asks how, once perspective had been accepted, it would be possible to paint a landscape without receding planes with distant objects in prospect and closer objects, larger and darker because of more apparent shadowing, appearing as refuges. Finally, Appleton is taken to task for failing to consider Islamic, Oriental, or even modern abstract art as data for his universalist theme.

A tendency to go overboard is common to both theorists. We are not convinced, for example, that tourists flocking to medieval Italian hilltowns are doing so to satisfy aesthetic needs rooted in 'collective [genetic] encoding ... of archetypal themes, culminating in massive urban projections of fundamental symbolism' (Smith 1977: 52). Nor are we wholly sure that Brennan's law of shopping behaviour, (which demonstrates that perceived distances towards the centre of a city are 'shorter' than the same distances on roads leading out of the city) demonstrates the existence of 'a deep-seated psychological pull generated by experience and memory, mixed with deep-rooted tendencies which link us with our urban ancestors in the Mesopotamian or Indus valleys' (p. 106).

Neither theory is readily operationalized or tested. Work by Clamp and Powell (1982) failed 'either to support conclusively or to negate the central claim of prospect–refuge theory' (p. 8), although both Woodcock (1982) and Nasar et al. (1983) found the theory a little more promising and Heyligers (1981) found 'prospect–refuge symbolism applicable at various levels of abstraction' (p. 8). The theory has been pressed into service in art history and literary criticism, and in an extensive review of his reviewers and replicators Appleton (1984) reasserts his right to put forward bold, untested or untestable theories. Given the rigid mores of academe, he tellingly adds: 'Perhaps it is only we elderly academics, put out to grass, who can really afford to do these things' (p. 103).

Beyond hypothesis-testing, neither Appleton nor Smith appears to recognize the need to create a broader-based theory of aesthetic origins which, though rooted in genetic encoding, might take into account fundamental cultural differences in world-view and aesthetic appreciation of landscape. Nevertheless, Smith and Appleton have produced very readable, thought-provoking, visually-appealing books, brave attempts to engender a search for the origins of the aesthetic appreciation of landscape. Few others have so ventured. Indeed, as Appleton has noted, (1975b, repeated by Dearden in 1989), the whole field of environmental aesthetics remains stoutly empirical, with a huge superstructure of research erected above a vast theoretical vacuum. Only with the development of the concept of biophilia has a wider biologically-based theory begun to emerge (Kellert and Wilson 1993). In this volume both Ulrich (1993) and Heerwagen and Orians (1993) provide thorough accounts of recent research and theory on the genetic approach to aesthetics, explicitly linking these with the activist conservation ethic of biodiversity. Nevertheless, this approach remains essentially speculative and empirical work must perforce be the baseline as we turn our attention to a consideration of the senses.

1.4 THE SENSES

More than eighty per cent of our sensory input is visual (Rock and Harris 1967). Psychologists, urban designers, landscape architects, and advertisers all stress vision as the chief mode of knowing about the world. So much so, indeed, that when we use the term perception we almost always mean visual perception. This myopia stems from our cultural prejudices and values (Avocat 1982), and from the ease with which we can study or control vision in comparison with the other senses. Yet the emphasis on vision seems rather quantitative; we have little information on the qualitative importance of other perceptual modes (Howes 1991).

There appear to be two basic modes of perception (Schachtel 1959). Autocentric (subject-centred) senses combine sensory quality and pleasure; the concern here is how people feel. In contrast, allocentric (object-centred) senses are concerned with objectification and knowledge; these senses involve attention and directionality. Vision, except colour perception to some extent, is chiefly allocentric. Speech sounds are allocentric, whereas most sounds, as with all the other senses, are autocentric. Table 1.4 contrasts these two modes of perception.

It is notable that while children are basically autocentric, in Western cultures they learn to develop allocentric modes until these become dominant. This is not true of other cultures, for many of these prize the autocentric, as with tactility (Japan), smell (Oceania) and kinaesthesia (Africa). Among Western adults, in contrast, vision is dominant and preferred, while autocentric senses such as smell are atrophied.

It is perhaps significant that Western people prize the cool, detached, intellectual sense which effectively distances them from their environment. With such distancing it becomes all too easy to regard nature and environment as a series of objects worthy only of disregard or exploitation.

It is necessary, therefore, to devote some attention to the chief characteristics of the four senses of most value in the interpretation of environments.

Table 1.4 Autocentric and allocentric senses

Autocentric	Allocentric (chiefly vision)
'hot'	'cool'
physical	intellectual
primitive	sophisticated
expressed *de novo* on each occasion	easier to recall and to communicate
more 'actual'	more detached
close range	distanced
basic in children	develop with maturation

Vision

As a physical system, human vision can hardly be surpassed in many respects by any piece of technical hardware. The eye has an amazing ability to distinguish differences in distance, in light intensity, and in spectral qualities (colour). Vision is the dominant sense in humans, providing far more information than all the other senses combined. Its relative dominance varies with culture. In Western cultures, however, it is clear that, in comparison with the other senses, vision is active, cool, and distant.

Vision is not pictorial (Rapoport 1977); rather, it is active and searching. We look; smells and sounds come to us. Orientation in space is chiefly achieved visually except in cultures such as the Inuit (Carpenter 1973). Visual perception relies on space, distance, light quality, colour, shape, textural and contrast gradients, and the like. It is a highly complex phenomenon.

When vision is experimentally placed in conflict with other senses it inevitably turns out to be the dominant sense (Rock and Harris 1967). Studies of aesthetic quality judgements in the field using composite visual, auditory and other scales have demonstrated that composite perceptions are heavily weighted toward the visual. The perceived environment, then, is largely a visual one, and most science is 'eye science' (Cunningham 1975).

This overwhelming reliance on vision has its problems. We do not have eyes in the back of our heads; our arc of vision involves only what lies before us, with gradually diminishing acuity to the sides via peripheral vision. Because what we see is 'out there', and we are physically unable to see what is 'right under our noses', there is inevitably a physical and psychological distance between ourselves and what we observe.

Further, the eyes read 'errors' into the brain. Those which are of importance to environmental aesthetics include the fact that distant objects appear smaller to us; we do not have telescopic vision. Bright-coloured objects appear larger than they really are. Perspective can be a problem unless we have enough background clues to the distance and height of objects. Optical illusion tests suggest that human vision exaggerates the vertical dimension, so that mountains appear to 'tower' when instruments inform us that they do not. Finally, the environment itself can increase 'error readings', as when in a thick, misty atmosphere objects loom larger or when suspended particulates enhance the warm colours of sunsets or typically induce us to see distant mountains as blue.

Colour is the exception to the tendency of vision to remain a distanced, detached sense, and colour vision thus has some similarities to the non-visual, more emotive senses. Like music, colour clearly registers emotional meaning in the observer. Typically, semantic differential tests record a spectrum from strong/active (deep blue, red) to weak/passive (grey-blue, pink) qualities. In general, Western cultures find red hues to be more active, stronger, and warmer than blues. On 'liking' or 'pleasantness' scales, however, the most liked colours in capitalist industrialized nations seem to be blues and purples, declining through reds to least liked yellows and greens (daffodil colours, ironically).

Psychophysical tests show that, given equal brightness, reds arouse people much more than do blues. American females, when given the choice of clothing colours of the same hue in several saturations (e.g. vivid red through pastels to pale pinkish tints), were found to choose shades according to personality rather than complexion and hair colour. Typically, red-choosers were outgoing and active, whereas pink-choosers were more submissive, detached, passive, and quiet.

Our ability to distinguish colours, like our binocular stereoscopic vision, is shared with relatively few other animals. It has, therefore, considerable importance environmentally in terms of emotion, sensation, mental states, and symbolism. Colour in traffic signals, interior decoration, clothing, and the symbolism of political party insignia and national flags, has significant meaning for us. Colour associations are all-important; most of us would reject blue food.

The deliberate use of colour is extremely familiar to urban dwellers and needs little elaboration here (Foote 1983). Much less is known about the colour properties of natural or agricultural landscapes, partly because landscape colours are difficult to classify and are extremely ephemeral. Nevertheless, Wood (1989) has made an attempt to regionalize landscape coloration (Figures 1.5, 1.6) and has explained how landscape colours change after human interference with natural vegetation cover. Typically, mono-culture farming, irrigation, and forestry decrease the colour variety of landscapes in favour of extensive monochromes. Human settlements provide high contrast through the extensive use of abiotic colours.

Clearly, it would be possible to discourse interminably on colour in the landscape, let alone the general properties of vision. It is far more useful, I think, to take vision for granted, as 'ground' in the figure-ground equation perhaps, and concentrate more fully on the three senses which, although accounting for perhaps no more than 10 per cent of our sensory input, are extremely important in terms of emotional, and thus aesthetic, impacts.

Sound

Other than vision, sound is the only sense extensively studied in environ-mental aesthetics. It contrasts strongly with vision in many ways. Unlike visual space, which is sectoral, acoustic space is non-locational, spherical, and all-surrounding. It has no obvious boundaries, and, in contrast with vision, tends to emphasize space itself rather than objects in space. Sounds, compared with things seen, are more transitory, more fluid, more unfocused, more lacking in context, less precise in terms of orientation and localization, and less capturable.

Audition is a fairly passive sense; one cannot close one's earlids. Sound, therefore, is ubiquitous; there is no end to traffic roar, building and machine hum, the rustling of leaves. Sudden silence can be extremely disconcerting. Some city-dwelling visitors are extremely disquieted(!) by the deep silence of the island on which I write this.

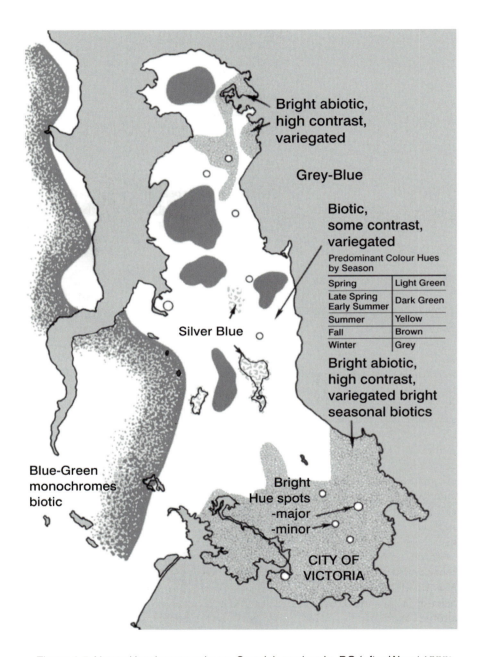

Bright abiotic,
high contrast,
variegated

Grey-Blue

Biotic,
some contrast,
variegated

Predominant Colour Hues
by Season

Spring	Light Green
Late Spring Early Summer	Dark Green
Summer	Yellow
Fall	Brown
Winter	Grey

Silver Blue

Bright abiotic,
high contrast,
variegated bright
seasonal biotics

Blue-Green
monochromes
biotic

Bright
Hue spots
-major
-minor

CITY OF
VICTORIA

Figure 1.5 Normal landscape colours, Saanich peninsula, BC (after Wood 1989)

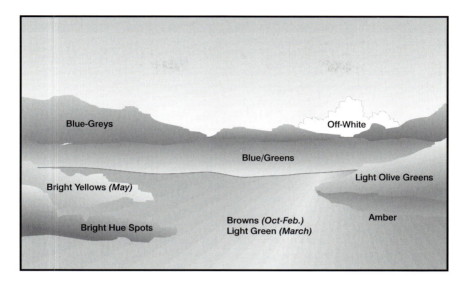

Figure 1.6 Colour dominants, southern Vancouver Island, BC (after Wood 1989)

The human sense of hearing is not very acute, ranging from sixteen to twenty cycles per second. Below this, heartbeats can be detected; at or above, bats squeak their nocturnal sonar. There is some evidence, too, that auditory acuity may be in decline in noisy urban-industrial societies. Even if the same musical instruments are used, it is unlikely that we hear what Mozart heard.

This reference to music introduces the observation that although, compared to vision, sound perception is information-poor, it is clearly exceptionally emotion-rich. We are strongly aroused by screams, music, thunder; we are soothed by the sounds of water, leaves, wind in the grass. Infants are especially sensitive to the pleasant–unpleasant range of sounds, and the development of deafness in adults may lead to loneliness, depression, and even paranoia. Loss of hearing reduces our sense of the progression of time, and contracts our sense of space, for we can normally hear further than we can see. Sound provides dynamism and a sense of reality. Rhythmic sounds such as music and waves reproduce the basic pulses of life – heartbeat and breathing – and are important contributors to well-being.

In environmental terms, soundscape is an important component of our sensory environment. I have discussed this in detail elsewhere (Porteous and Mastin 1985) and complementary approaches are to be found in R. M. Schafer (1977) and Truax (1984). It remains to note that soundscape research almost invariably finds that traffic noise is the most ubiquitous 'ground' (or 'keynote') sound in Western urban environments, and that it seems to be gradually masking 'figure' sounds which are usually perceived by citizens to be more pleasant or more important. Traffic roar increasingly drowns the sounds of nature as well as public music and informational sounds such as foghorns and even sirens. Late twentieth-century urbanites, therefore, in their demand for noisy technological junk, are paying the high price of growing sensory deprivation.

In terms of urban design, it would not be difficult to manipulate acoustic space in cities as well as in buildings (Scheer 1979). But one condition for an auditorily-pleasant environment is a reduction in the noise of machine and motor sounds. The serious underfunding of scientific and technical work on this problem is typical of our culture's lack of concern for environmental aesthetics and our apparent willingness to undergo physiological deformation and the consequent loss of sensory pleasure that comes with declining acuity. In a noisy urban world, perhaps, a degree of deafness may even be adaptive.

Smell

As with sound, the human sense of smell is not well-developed. Dogs, for example, have an olfactory acuity about one hundred times greater. Non-human primates rely heavily on smell for safety and survival. Even for humans, the sense of smell is of vital practical importance. The smell of food may increase its appeal or warn us against tasting what might be poisonous or nauseating. The smells of home, of persons, of pets, of cuddly toys or of a 'security blanket' are reassuring and deepen our sense of attachment to environment and society. Smell also has its part to play in sexual arousal, and not only for those moths that can apparently smell their sexual partners at distances of up to seven miles. Olfaction, then, is an important component in the satisfaction of our physiological, safety, and affiliation needs. It is basic to our well-being.

Environmentally, smell is even more information-poor and emotion-rich than sound. Smellscape is an enveloping, unstructured, often directionless space; the smell-world is 'diffuse, inchoate, transient, and emotional. Odours arouse feelings of pleasure, well-being, nostalgia, affection, and revulsion. They are direct, specific, ungeneralizable experiences – ends in themselves' (Tuan 1982: 117). Plugged directly into the limbic brain (Gloor 1978), olfactory receptors may unleash the primitive in us.

The child's environment is far more richly olfactory than that of the adult. There is evidence to suggest that olfactory acuity, or at least the amount of attention paid to smells, diminishes after puberty. Yet most adults can distinguish artificial from natural objects by smell, and a select band of perfume, wine and tea experts have a wonderfully developed olfactory acuity. It is possible to identify whole cities and landscape regions in terms of smell, from Hershey, Pennsylvania (chocolate) through Tadcaster, England (brewing) and Uji, Japan (green tea) to the sour gas smells of oilfields and the sulphite smell of pulpmills in Western Canada.

Old-fashioned cheese, hardware, tobacco or chocolate stores can be an olfactory delight. Yet North American cities have become increasingly smell-impoverished. Unlike their counterparts in Mediterranean Europe, North Americans rarely relish smells. Most environmental smells have been designed out of existence by means of personal and public deodorants, air conditioning, air cleaners, and plastic packaging. The keynote smell in our cities has become, like the keynote sound, that of the motor vehicle. Sensory impoverishment rules in a world selectively 'sanitized for your protection'.

Very little has been written on environmental smells, though the reader may be referred to my earlier work (Porteous 1990) for an overview. In practical terms, zoning laws have long recognized the problem of smell in their segregation of noxious land uses such as slaughterhouses, glue plants, and chemical factories. On the positive side, tactile museums for the blind have been matched by the construction of odoriferous gardens which provide the visually-handicapped with rich olfactory sensations, give directional information, and confirm the passage of the seasons. Commercially, odour is an important tool in marketing.

Environmental odour, however, has generally been considered only as a problem. Just as soundscape studies are dominated by noise research, so the investigation of smellscapes is almost wholly devoted to odour pollution. Indoor air quality and traffic odours have aroused considerable interest; about half of all complaints about air pollution involve smells. Odour pollution is now rapidly increasing in some rural areas with the spread of factory farming, which produces large quantities of noisome animal waste. Odour control technology includes odour dilution through heightening emission smokestacks, the use of scrubbers or combustion, and the masking of unpleasant by pleasanter smells.

As with soundscape, smellscape is an emotive environment, not an intellectual one, and as such deserves to be cherished. Life in future blandscapes will be severely impoverished if negative smells are annihilated but little effort is made to promote pleasant environmental odours. Asked what they missed during their record-making 211 days in space in 1982, Soviet cosmonauts replied: 'the smell of flowers, city noises, city smells ...' (Berezovoy 1983).

Tactility

Tactility, the haptic sense, produces a touchscape. The skin is our largest sensory organ, and it is extremely sensitive. The application to the skin of a pressure of the order of 0.04 ergs of energy can be detected, which is 10^8 times less than the minimum energy level detectable by ear or eye (Bouman 1979). With our fingertips we can pick out grooves etched in glass to a depth of only 0.0025 of an inch.

Even in the most sophisticated sensory deprivation experiments, the sense of touch cannot be masked; we cannot turn it off. Sensation-rich and information-poor, it is the most primitive and sensuous of all senses. Touch is vital for well-being. Children, pets, and sexual partners depend on touch. We gain information as well as pleasure by handling things, and for some, as with Doubting Thomas, touching, rather than seeing, is believing. There is much pleasure to be had from cool sheets, warm bodies, magnolia petals, and the feel of liquids. We are always in touch with our environment; at this moment I can feel paper, pen, desk, a sheepskin-covered seat, the texture of Berber carpet beneath bare feet, the cling of clothes to my body. Indeed touch, like seeing, is a basic language idiom. Metaphorically, we rarely wish to be 'out of touch'.

Yet touching with the fingers demands effort. Professional cloth-feelers apart, much of our experience of texture comes not via the hands but through the feet. Even if shoes are worn it is possible to distinguish qualities (soft, hard, smooth, rough) and types (grass, pebbles, sand, boardwalk) of surface. The textural changes of soft mats, polished wood, grained wood, and heavy paper are characteristic of traditional Japanese dwellings. In contrast, there is an increasing predominance of concrete and asphalt in modern cities. Textural experience, just as with sound and smell, is thereby eliminated or severely reduced. An increased use of textured sidewalks (gravel, tile, metal, glass, brick, wood) and Portuguese-style mosaics would greatly enhance the sensuous pleasure of city walking.

Tactility is clearly related to the subsidiary senses of kinaesthesia and the sense of temperature and air movement. The former involves the speed of bodily movement, the sharpness of angles and curves, the rate of directional change, declivity, slipperiness, and changes in body orientation (Figure 1.7). An appreciation of air temperature is the property of nerve endings which can detect changes of a few tenths of a degree Centigrade. The blind can distinguish up to a dozen air speeds (Rapoport 1977). These senses become important environmentally chiefly in terms of contrast, as when one moves from the blinding light of a street to the dark coolness of alley or church, from hot, dry city centre to cooler, moister sea-coast or park, or from the windy to the protected side of a tall building.

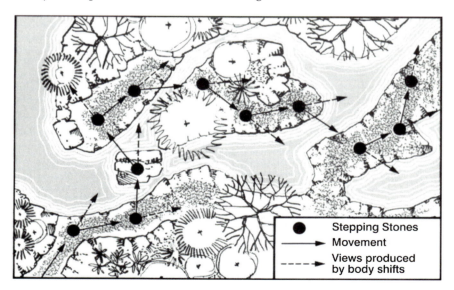

Figure 1.7 Stepping-stones: the link between visual and kinaesthetic experience

Figure 1.8 Reduction of rich sensory experience. Above, medieval market still operates in Selby, Yorkshire. Below, housing project anywhere

Figure 1.9 Visual aesthetics only; other senses elude representation: Saturna
Island, BC

Conclusion

Environmentally, places can readily be described by the characteristic 'mix' of sense perceptions available to the average able-bodied member of the public. The identity of places is multisensory, but in some cases one or more sense perceptions may be dominant. Sensory interaction is vital; people who hear more, for example, also see more (Southworth 1969). Kinaesthesia (motion by foot or vehicle) is the sense which most readily helps integrate the other senses over time.

Unfortunately, in modern environments both sensory variety and quality are rapidly being reduced to monotony through deliberate manipulation and via the heavy dominance of the motor vehicle (Figure 1.8). Attention to sensory quality coupled with sensitive urban design could alleviate this problem. It is more likely, however, that urbanites will continue to regard cities as sensory blackspots and seek sensory pleasure chiefly indoors, in private gardens, public parks, and nonurban rural and wilderness areas. And it is only because of the sheer dearth of research on environmental smell (Winter 1978, Corbin 1986, Le Guérer 1988) sound (other than noise) (Corbin 1994) and touch in the landscape that this book will be devoted chiefly to visual aesthetics (Figure 1.9). The reader who wishes to delve further into the sensory is referred to the recent work of lawyers (Hibbitts 1992), anthropologists (Howes 1991), historians (Classen 1993), geographers (Porteous 1990, Tuan 1993, Rodaway 1994), psychologists (Rivlin and Gravelle 1984) and popularizers such as Ackerman (1990) and Tiger (1992).

2

HUMANISTS

Convolvulus

What would the world be, once bereft
Of wet and wildness? Let them be left,
O let them be left, wildness and wet;
Long live the weeds and the wilderness yet.
                              Gerard Manley Hopkins

It's a false but popular truism that beauty is in the eye of the beholder. In other words, *de gustibus non est disputandum*, *chacun à son goût*, you can't dispute a person's taste. All is relative in a world of conformist individualism. Hence 'anything goes', with the result that we live in an ever more ugly world where to complain of another's lack of taste is to be branded an elitist. This chapter refutes this pernicious pseudo-democratic doctrine by appealing to history and sweet reason.

But our society purports to reject any possibility of learning from the past. 'That's history', we say of a past event, as if the past had no influence on present or future behaviour. Such phrases symbolize the profound ignorance that prevails in our culture, and it is on the basis of this ignorance that we attempt to build a myth of the eternal now. But as T.S. Eliot (1974: 189) tells us, the future contains both past and present, and is embedded in the past. On a personal level, one's behaviour is based on one's character, and this in turn derives from patterns formed before we can even remember forming them. Similarly, one's individual aesthetic appreciation is strongly influenced by one's cultural background, which one cannot help but absorb unconsciously as one grows up.

In aesthetics, then, social consciousness takes precedence over individual consciousness, and thus 'such hopelessly subjective formulas as "beauty is in the eye of the beholder" simply will not do' (Frye 1991: 31).

This chapter therefore begins by outlining the qualitative humanist approach to the understanding of nature, landscape and scenery (2.1). It then uncovers the history of environmental aesthetics in the Western world since medieval times, pausing where necessary to relate historic aesthetic concepts to current aesthetic beliefs (2.2). This historical outline is given spatial depth by complementary discussions of our changing tastes for landscape types such as mountains, wilderness, gardens and cities (2.3), and of national contrasts in current landscape taste (2.4). I conclude that the quantitative and applied pursuits of experimentalists, activists, and planners are inextricably grounded in the ideas generated by humanists.

2.1 THE HUMANIST APPROACH

Of the four approaches to environmental aesthetics, the humanist approach tends to be the most contemplative. Humanists are critical observers of human nature, landscape, and interactions between the two. Their concern is with the life of the mind; with the contemplation, rather than the manipulation, of environments and human behaviour. The approach is nonpositivist, hermeneutic, idiographic, sometimes explicitly phenomenological or existentialist. Personal and group experience, intuition, and inductive reasoning are stressed. In sharp contrast with mainstream scientific approaches, human values are at the forefront.

Because of their interests in the cultural roots of aesthetic behaviour, humanists have a strong leaning toward the past. Much emphasis is placed on tracing historic trends in landscape taste. It is from this work that we have come to appreciate the importance and essential temporality of 'currents of taste', such as the fall and rise through the last millennium of Western appreciation of mountains and wilderness.

Consensual experience is understood through the assessment of novels, poetry, paintings, diaries and other personalist art forms. This results in a tendency to extrapolate the aesthetic feelings of the literate elite to society at large. We know, for example, a great deal about the aesthetic theories of well-to-do eighteenth-century Englishmen, and can still trace their expression in concrete form in field, park, and town. We have very little comparable information about women or about the mass of the population who were compelled to live in and among these contrived landscapes. Only with the analysis of folklore or other plebeian art forms could any account be made of the tastes of the majority (Bunkse 1978, Silk 1984). Emphasis on elite tastes, however, is justified in that tastes tend to 'trickle down' through the centuries from elite groups to the population at large. Aristocratic eighteenth-century taste in gardening, for example, is reflected today in the vast landscapes of suburbia and in city parks.

Nevertheless, because of this elitist frame of reference, humanist work has sometimes concentrated upon the exceptional in landscape. Existing literature tells us a great deal about the aesthetics of Georgian cities, eighteenth-century country estates, Italian belltowers and French bastides, but much less about suburbs, apartment zones, or inner-city neighbourhoods. In both Relph's *Place and Placelessness* (1976) and Smith's *Syntax of Cities* (1977) the examples chosen to illustrate authenticity and high aesthetic value are almost universally pre-modern. In general, there seems to be an anti-modern, anti-urban bias, with a preference for architecture and landscapes created under elitist or tyrannical forms of government. A number of geographers have attempted to correct this emphasis by concentrating their work on 'ordinary landscapes' and the 'everyday urban world' (Meinig 1979, Porteous 1977).

Beyond history and culture, a significant trend among humanists is to plumb the universals of experience in space and time. Important dialectics emerging from this work include space and place, home and abroad, inside and outside, front and back, surface and depth (Tuan 1974, 1989). The positivist construct of undifferentiated, objective space is contrasted with differentiated subjective place, pregnant with meaning, laden with emotion, and highly expressive of the intentions of its inhabitants. From such humanist work, some of it over two thousand years old, emerge useful concepts which can be tested by experimentalists, absorbed by activists, and applied by planners.

Finally, much of the work produced by this group of aestheticians is eminently readable. Indeed, the written product of these investigations is as much a literary genre as a scholarly attempt to discover principles underlying our appreciation of landscape. Above all, contemplation and understanding are regarded as ends in themselves, or essential adjuncts of a richer personal life, while immediate utility is at a discount. In terms of practicality, however, a humanist aesthetician might argue that before one can understand a landscape one must first lay bare the cultural assumptions that underlie that landscape.

From Nature to Landscape

Environmental aesthetics covers both natural, contrived, and built environments. We have little trouble conceptualizing the built environments of cities or the contrived, semi-natural landscapes of rural areas. Nature, however, poses a problem: what is the nature of nature?

Nature is a highly complex concept (Tuan 1977, Relph 1981, Gold 1984, Olwig 1993, Schama 1995). It is generally agreed that nature exists only as a social product, an artifact of art and discourse. There is no way to specify the attributes of nature 'other than through the conventions and categories of human experience and thought' (Goodman, quoted in Tuan 1977: 28). While environment may be a more neutral term, expressing what is 'out there' in relation to what is 'in here' (the human brain or, more commonly, the body or person), 'nature has no meaning other than what we bring to it. Nature is inarticulate'. Consequently, there are as many 'natures' as there are cultures or schools of thought or art within those cultures.

In Western culture we appreciate nature in two major ways. Knowing is objective, detached, scientific; nature is an object for the subjective self. In contrast, feeling is much more ambiguous, and it is harder to distinguish subject from object, for the two become connected not by rationality, as in knowing, but by affective emotion. 'Art', says Tuan (1977: 32), 'is an image of feeling.'

It is clear that art may imitate nature. The notion that nature imitates art, espoused by Oscar Wilde in the late nineteenth century, was then considered shocking. But it is clear that artistic images are used by people not only to reshape nature but also to 'see' nature. Only with the efforts of the Group of Seven painters in Ontario and Emily Carr in British Columbia did Canadians come to see the national set of natural landscapes in a more positive aesthetic sense.

Landscape is in part a subset of nature, and our appreciation for it seems rather recent. Developed from the Dutch concept of 'landschap' which emerged *c.* 1600 (Tuan 1977, Relph 1981), landscape initially meant merely the background of a portrait or a rather commonplace view of farms and fields (Stilgoe 1982). Only when later translated to Britain by gentlemen, artists, and gardeners did the concept develop more noble attributes as a visually pleasing prospect whether on the ground or on canvas. In eighteenth-century Britain landscape became 'an aesthetic concept that required leisure and training to appreciate' (Relph 1981: 23).

The subsequent loss of this leisure and training has meant that, in twentieth-century terms, landscape has become almost indistinguishable from scenery. Originally, landscape delighted all of the senses and was best appreciated kinaesthetically, by moving through it (see Chapter 5.2). Scenery, in contrast, is a wholly visual construct (Sautter 1979), a relatively simple notion deriving from the furniture and painted backcloths of stage decor. Scenery is two-dimensional; it has no depth; it is a mere backdrop for human action. We have little involvement with it, and its essential falsity is seen in such phrases as 'behind the scenes'. The modern equation of landscape with scenery, therefore, can be seen as an act of bad faith, tending to reduce meaningful aesthetic interactions between person and environment.

It becomes easier, then, to say 'it's only scenery' when we wish to destroy landscape for profit. Here Meinig's (1979) concept of 'the beholding eye' places aesthetic appreciation in context, as only one of at least ten modes of viewing the same scene. The concept assumes that landscape is a form of communication as well as the result of human desires (Sanguin 1981). Three spiritual or ideological views are noted first. Those who see a scene as 'nature', wherein humankind's feeble works are overwhelmed by the ageless power of natural forces, are contrasted with those who see 'artifact', where people conquer nature and create humanized landscapes. A more balanced view sees the scene as 'habitat', where humankind and nature work in harmony to produce an essentially humanized, but not dominated, earth.

Scientific views come next. The fourth observer sees the scene as a 'system, a dynamic equilibrium of interacting forces', whether physiographic or economic. The applied scientist, next in line, appreciates the system but observes the scene as essentially a 'problem' which can be resolved by the application of technical expertise.

Further contrasts emerge with observers six and seven. One of these expresses the extremely common view of environment as a resource, as a source of wealth. Even views are commodified and sold as waterfront property, hillside lots, and penthouses. The next observer agrees, but stresses that the landscape chiefly reflects the prevailing cultural ideology, which in North America, at least, centres upon wealth-production. The landscape ideologist sees the scene as philosophy translated into tangibility. She reads the landscape as a text, and a typical American landscape, for example, may be seen as reflecting a dominant value-system which lauds freedom, individualism, power and progress.

Less radically, the scene can be viewed as 'history', as a layered palimpsest of what has gone before; landscape as process and accumulation. Allied to this is the view of landscape as 'place', a unique entity, a repository of affections of both inhabitants and visitors. Finally, the tenth observer believes the 'artistic quality' or 'aesthetics' of the scene to be of major importance. Here aesthetics varies from the 'pure', where the scene is appreciated as colour, line, mass, texture, and balance, to the 'less pure', which involves the notion of landscape meanings and symbols.

Clearly, these ten views resolve into a number of clusters. We would expect that those regarding the scene as 'artifacts', 'wealth', 'system', and 'problem' might have more in common with each other than with those who regard it as 'nature', 'habitat', 'place' or the aesthetics of form, line, and colour.

To expand this latter concept of landscape as aesthetic, Thayer's (1976) notion of visual ecology is helpful. Thayer, speaking chiefly for landscape architects, suggests that landscape has at least five levels of significance. Three of these have an aesthetic component.

The first level is the concrete or presentational level, a pure aesthetic surface of light, line, form, mass, colour, and texture. In the second, or associative level, we find this aesthetic surface has become a tree with a function or use; it gives shade. The fact that this tree may calm, affright, or interest us brings us to the third, emotional or affective, level. When we measure the tree against some value system we reach level four, the rational or symbolic level. The sudden rage in 1989 for public sculptures of plastic, bronze, and silver trees in Victoria, British Columbia, can be seen either as a vibrant symbol of the province's premier economic base, the timber industry, or as a sadly ironic reminder of the imminent extinction of old-growth forests in Western Canada. Finally, at the behavioural or activating level we are provoked to action; we sit under the tree and enjoy its beauty, or we chop it down to make comic books and toilet paper.

Lewis's (1979) 'axioms for reading the landscape' echoes the con-ceptualizations of Meinig, Thayer, and others. In terms of environmental aesthetics, his most useful ideas confirm the need to understand historic tastes in order to appreciate current landscapes and the need to pay attention to the common or ordinary landscapes in which most of us live. Lewis's axiom that landscape reflects culture generates a 'corollary of taste'. So much of what we regard as normal or ordinary in landscape actually reflects our rather specialized tastes. To give taste a heavier load of meaning we can illustrate this notion by observing that had Americans developed a taste for mutton and shunned dairy products, the 'beef landscapes' of the Midwest and the 'dairy landscapes' of the Northeast would never have emerged. Similarly, had we rejected our historically-derived taste for grass lawns in favour of artificial surfaces or natural growth, then the fronts of our houses might well be graced by grey concrete slabs or ecologically-sound forests of 'weeds'. This corollary of taste is all-important. Too often we unthinkingly accept that our landscapes are 'normal' while those of others are the product of rather unusual tastes or merely a temporary stage in a teleological development toward a North American norm.

That the North American norm is a historically-derived social product will be discovered in the pages which follow. It is time, then, to turn to a historical overview of the development of landscape tastes in the Western world. This overview will be approached in three distinct ways, each of which overlaps to create a three-dimensional view of landscape tastes through space and time. First, general currents of landscape taste through time are analysed. Second, these changes in taste are interpreted in concrete terms in the shape of the major elements of landscape, such as city, countryside, and wilderness. Third, this two-dimensional view is given depth through the analysis of variations in Western landscape tastes among English-speaking nations. Taken together, these three approaches should generate a synthesis of some value in the appreciation of the landscapes with which the reader may be personally acquainted.

2.2 A HISTORY OF LANDSCAPE TASTE

In some ways the Western development of art from medieval symbolism to post-Renaissance naturalism was prefigured in pre-Christian times. Early art is idealist in the sense that it depicts the idea of an object, not the object as seen. In other words, Mesopotamian, Egyptian, and Cretan art was conceptual rather than representational. Yet by the Classical period, say from 400BC onwards, a degree of naturalism had arisen. The purpose of such work was chiefly decorative. As classical landscape painting was chiefly used as a form of wallpaper, few examples survive.

The importance of classical Greek concepts to modern aesthetics lies chiefly in the legacy of ideas which, after lying dormant for a thousand years, were resurrected in the Renaissance and came to fruition in eighteenth-century Western Europe. Chief among these ideas is the scientific outlook. Concepts of science and logic were to be revived and become major factors in European aesthetic world views in the seventeenth and eighteenth centuries. The notions that there was some form of religious order in nature, and yet that humankind was the centre of the universe and inseparable from it, were to be revived as major tenets of the Romantic movement of the early nineteenth century, which saw no problem in the reconciliation of science and religion. The loss of the Greek conception of time, as essentially cyclic rather than linear, has been in part responsible for the development of the destructive modern notion of progress.

The medieval world

With the victory of Christianity in the fourth-century Roman Empire, a vast change took place in the Western conception of environment. First, humans were no longer seen as inseparable from nature. Indeed, strict interpretation of Genesis suggests that humans are the possessors of nature. Naming, one of Adam's first acts, is a possessive act. Adam's descendants, having lost Eden, were given dominion over all living things and told to 'go forth and multiply'. Man, after all, was created in the image of God, not in the image of nature. Hence the modern concept that humankind and nature are dualistic, two separate entities, and hence the loss of human respect for nature that has had such serious ecological consequences.

Second, having lost respect for nature, humans were now free to exploit it without fear of punishment. God, who notices the fall of a sparrow, gives his covenant not to the sparrow, but to humans. Earlier ideas that the earth is imbued with spiritual power, or has religious value in itself, became lost except in marginal, non-Christian, 'primitive' cultures. The earth, according to scripture, was created by God for human benefit; the earth is a resource. Since the loss of Eden, and with it the joys of sensuous experience, the earth was no longer to be freely enjoyed, but to be exploited under God's guidance. Here are the well-springs of modernity.

Third, with Christianity came the notion of linear time, of beginnings and endings, of both time and humankind 'going somewhere'. For humans, the somewhere was heaven or hell, for which the earth was merely a waiting room or vestibule. Time as an arrow, desacralized, has become the modern norm, the basis of 'progress'.

The consequences of these changes in world view were overwhelmingly significant. Humans lost any sense of being part of a natural process, any sense of 'environmental humility'. Cyclical, seasonal time, although still of obvious importance in an agrarian world, became subordinated to a sacred linear progress, symbolized in our time by the assembly line. And people came to look upon nature in a merely utilitarian, instrumental way, as a resource.

Medieval Christendom, then, had a transcendental view of the earth, which came to be seen largely in symbolic terms. Indeed, it was thought that the world had been created as a set of symbols for our spiritual edification, as an assistance towards heaven. As with the Platonist Greeks, but with a teleological twist, the idea of a thing became more important than the thing itself. The way to appreciate the world was not via the senses, but as allegory. In Bunyan's *Pilgrim's Progress* we do not encounter wetlands or sylvan vales, but Sloughs of Despond and Valleys of the Shadow of Death.

Indeed, the senses were seen as debased. The strict monastic view was that nature itself was sinful. St. Anselm (*fl.* twelfth century) felt that things were harmful in proportion to the number of senses they had the power to delight. Laymen, perhaps, would have not thought it wrong to enjoy nature, but the medieval labourer 'would simply have said that nature was not enjoyable' (Clark 1956: 18). More positive lay attitudes to nature emerged only in late medieval times in calendars, church sculpture, and the poetry of Chaucer.

The result of these attitudes is the remarkable absence of medieval landscape painting or poetry. Most painting was highly symbolic or allegorical, and almost always religious. Byzantine art became so rigidified that icon-painting styles changed little from the fourth century to the fourteenth. Where landscape appears in Western art, it is as background to a symbolic narrative involving figures of saints or the holy family (Figure 2.1). Nevertheless, it is possible to gain a general idea of medieval attitudes to nature.

Nature, to medieval people, was disturbing, vast, and fearful. The ordinary landscape of the medieval person consisted of villages and fields, the larger houses surrounded by walled gardens. Beyond the walls, beyond the village, lay enormous swathes of forest. Grimm's fairy tales, Dante's *Inferno*, and innumerable folk tales convince us that, notably in Central Europe, forests were regarded with horror and dismay. Silent, dark, cold, airless, full of unknowns, forests are the abode of beasts and bandits, and only partially penetrable while hunting. According to Clark (1956: 18) 'there is ... something in the character of great forests which is foreign, appalling, and utterly inimical to intruding life'. Little wonder that the medieval peasant's chief role in landscape change was to destroy forests in favour of farmland.

Beyond the forests lie mountains; dark, forbidding, cold, unknown and, despite Moses and the Greeks, now regarded as unknowable, and certainly no longer the seats of the gods. There was no concept of climbing them. Indeed, the first 'modern' man to climb a mountain for its own sake was the poet Petrarch in the late fourteenth century. But hardly had the poet begun to enjoy his superb view of the Alps, the Rhône, and the Mediterranean,

Figure 2.1 Landscape as background: Lorenzo di Credi (1457–1537) *Virgin and Child* (reproduced by courtesy of the Trustees, The National Gallery, London)

when he was moved to open St. Augustine's *Confessions* only to find the passage:

> And men go about to wonder at the heights of mountains, and the mighty waves of the sea, and the wide sweeps of rivers, and the circuit of the ocean, and the revolution of the stars, but themselves they consider not.

Abashed and ashamed, Plutarch reported that 'I turned my inward eye upon myself.' Nature contemplation clearly lays open the mind to dangerous and unholy thoughts.

In medieval art, then, we find little landscape. Art was didactic; even if medieval artists had had the skills to paint realistically, they probably would not have done so. The environmental emphasis in medieval art lies on the city, a place of order and virtue, perhaps a forerunner of the Heavenly City. The only obviously pleasant natural scenes are of gardens, but these again are highly ordered and geometrical. Medieval gardens and their Renaissance successors meant nature conquered, cut to shape, and almost 'man-made', nature boxed-in like a domestic pet (Figure 2.2).

Beyond the artificiality of city, garden, and village lay the unholy wilderness, full of mischief and evil, a place of dragons and sprites, beasts and unknowns, an empty place into which only wild men and saints ventured. Wilderness was the dominant medieval landscape. It was feared because of its genuine dangers, because of its unknown, symbolic dangers, and also because of its psychological dangers, for wilderness was seen as an outward pro-

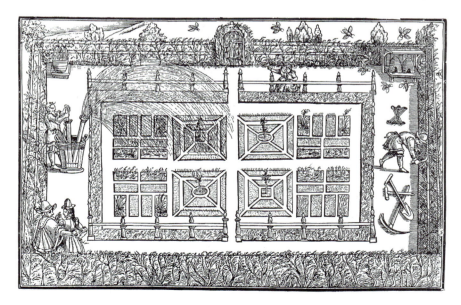

Figure 2.2 Medieval and Renaissance gardens; nature cut to shape. From T. Hill's *The Gardener's Labyrinth*, 1586

jection of the animalistic, demonic, dark side of human nature. This latter notion was certainly part of the Puritan baggage that came ashore in early New England.

Given this view of the natural world, it is not surprising that medieval monasteries were established in wild places – deserts, forests, islands, and moorlands – where there was nothing, at that time, to delight the senses.

The Renaissance

By late medieval times the fear of nature had diminished sufficiently to allow accurate observation and depiction of plants and birds. These were generally seen as isolated objects, however, and rarely put together so as to form the composite picture which we would call a landscape. Despite a few notable exceptions, such as Ambrogio Lorenzetti's fourteenth-century frescoes of *Good and Bad Government* in Siena (Cole 1980), the notion of landscape art came only with the Classical revivals of the Renaissance. John Ruskin, indeed, divided all Western art into two great categories: 'symbolic' art from the third to the fourteenth centuries; and 'imitative,' or naturalistic, art thereafter. After Roger Bacon's scientific revival of the thirteenth century the urge to observe, experiment, and record led to a growing feeling that nature was interesting in its own right, and that individual experience was at least as important as theological revelation.

Renaissance landscape art, however, was not monolithic. Indeed, two major schools contended, the Naturalistic school of Flanders and the Idealist (or Abstractionist) school of Italy. Not until the nineteenth century would the former finally triumph.

The Italian Idealist school of painting was more truly Renaissance in style. The chief object of these painters was to resurrect the mythical pastoral Arcadia of the Latin poets, such as Virgil and Ovid, who themselves looked back to the supposed Golden Age of Greece (Olwig 1993). This was essentially normative painting, for the painters depicted not what a landscape actually looked like, but what it ought to look like. The ideal was classical pastoralism: tranquil, serene, and very civilized. Nature appeared as a pastoral symphony of tranquil water, grazing sheep, cropped sward, and simple country pleasures among pleasant groves and small classical temples. Nature was seen as a source of joy and contentment, scaled down wholly to human dimensions, improved, ordered, and tamed, and bent wholly to human needs. This formal Arcadianism came to fulfilment in the landscape improvements of English gentlemen in the eighteenth century.

The three great landscape painters of the school were Claude Lorrain, Nicolas Poussin, and Salvator Rosa. Claude painted the Mediterranean countryside through rosy Arcadian spectacles. Although he painted directly from nature, with much detailed observation, and with a spontaneous feel for natural beauty, the end results were inevitably tamed, park-like settings (Figure 2.3). Poussin, in turn, painted heroic landscapes, where nature was far more strictly ordered, usually via the imposition of classical architectural

Figure 2.3 Claude Lorrain (1600–82) *Landscape with Flight into Egypt*

Figure 2.4 Nicolas Poussin (1594–1665) *Landscape with Diogenes* (courtesy Musée du Louvre, Paris)

forms or the insertion of robed human figures (Figure 2.4). In sharp contrast, Salvator Rosa's landscapes were much more melodramatic, full of precipices, bandits, wolves, and shaggy fir-trees, a violence more attractive to early nineteenth-century Romantics than to the more urbane eighteenth century (Figure 2.5).

Figure 2.5 Salvator Rosa (1615–73) *Grotto with Waterfall.*
(Courtesy Soprintendenza per i Beni Artistici e Storici, Florence)

Figure 2.6 Aert van der Neer (1603–77) *River Landscape by Moonlight* (courtesy the Rijksmuseum, Amsterdam)

Table 2.1 The influence of Renaissance art on later periods

The Platonic World	The Aristotelian World
The Renaissance of the Classical	The Renaissance of the Scientific
17th century Idealism (Italy)	17th century Naturalism (Dutch)
18th century *regular* landscapes or parks (nature tamed)	18th century 'sublime' and 19th century *irregular* landscapes, romanticized (nature wild)
The modern lawn, the city, the suburb	The modern wilderness cult, tourism, conservation

The Naturalistic, or Dutch, school had a very different world-view. As early as the fifteenth century Dutch painters such as the van Eycks were painting nature as fact. But this promising development was soon swamped by Italian Idealist influences. Rees (1973) quotes Michelangelo's contemptuous opinion of the absence of classical values in contemporary Dutch art: 'stuffs, bricks and mortar, the grass of the field ... bridges and rivers which they call landscapes ... done without reason, symmetry, or proportion'.

Not until the later seventeenth century did the Dutch re-establish their naturalistic style in the works of van Ruysdael, Rubens, and Hobbema. The reasons for this re-emergence probably include an increasingly scientific view of the world which followed Dutch work in the contemporary Scientific Revolution, the growing importance of a developing bourgeoisie of down-to-earth merchants who cared little for classical ideas and wanted their paintings to reflect the world they were building, and the cessation of the wars of the Counter-Reformation, landscapes now being seen as symbols of peace and tranquillity.

And indeed Dutch landscapes are tranquil, dominated by sky and water (Figure 2.6). These landscapes of little towns, fields, windmills and ships are instantly recognizable. Their emphasis on the world as it is had a profound influence on nineteenth-century aesthetics. Table 2.1 above summarizes the great importance of Renaissance art in influencing the ideas of later periods.

The eighteenth century

The eighteenth century is very much the turning-point in any history of environmental aesthetics, for it is at this time that many of our current landscape tastes were originated or formalized. Indeed, we may hold eighteenth-century aesthetic theorists responsible for a great deal of the landscape in which we live today, including suburbs, lawns, city parks, and our predilection for scenery, tourism, and the outdoors.

During the eighteenth century the European landscape improvement movement became dominated by English theorists and practitioners. Increased urbanization and industrialization, the effects of the Scientific Revolution, improvements in travel, the decline of religion, all these were background to unprecedented change. Aristocratic and even bourgeois youths were taken on the Grand Tour to Europe, where they came into contact with Italian paintings, classical statuary, and larger, wilder, more asymmetrical landscapes than they had previously known.

This was also the time when aristocrats, politicians, the rich, and the leisured looked upon landscape appreciation, and therefore landscape reshaping, as a very suitable occupation for educated people. Landscape became an art form, and whole tracts of countryside were reshaped to fit contemporary theories of how landscapes should look. Eminent theorists produced learned tomes, practical gardeners became much sought-after experts, and some of our now-common notions were established. Greek ideas of the Golden Mean and Golden Section were resurrected, while landscapists were informed that 'Nature abhors a straight line' (William Kent, painter and gardener) and that the true 'line of beauty' was curvilinear or serpentine (Hogarth). Theorists felt able to categorize landscapes generally into three types, the beautiful, the sublime, and the picturesque.

The notion of the beautiful in landscape was derived from Persian, Biblical, and Greek sources. In essence, the beautiful landscape was the tamed agricultural landscape, whether arable, pastoral, garden or orchard. The Italian Idealists (Figures 2.3, 2.4) clearly produced beautiful landscapes, which were widely copied on the ground in eighteenth-century gardens and parks. From this concept derive the English notion of 'countryside' and the far more general preference for lawns and suburbs.

The pastoral may be regarded as a subset of the beautiful (Figure 2.7). Psalm 23 sets the scene:

Figure 2.7 The pastoral: sheep safely graze in the Yorkshire Wolds

> The Lord is my shepherd: I shall not want;
> He maketh me to lie down in green pastures;
> He leadeth me beside the still waters;
> He restoreth my soul.

Here is the tranquillity bestowed by nature, and especially by water, a tranquillity able to restore the souls of those retiring to country mansions from the hurly-burly of the city. The pastoral/beautiful was to become the dominant landscape in much of eastern and central England with the vast changes wrought by the enclosure movement and the laying out of landscape gardens. Much of the rural England that tourists flock to see is the product of a combination of eighteenth-century aesthetics and economics (Figure 2.8). Pastoral poetry was at its zenith. Nymphs and shepherds, with names like Amaryllis and Phyllis, capered to innocent pastoral music. Dreadfully dull, laboured poetry, in which each stanza had to have its classical reference, was the norm. Pope's awful first lines, at age sixteen, provide some insight into the provenance of the notion of the beautiful:

> First in these fields I try the sylvan strains
> Nor blush to sport on Windsor's blissful plains;
> Fair Thames flow gently from thy sacred spring
> While on thy banks Sicilian muses sing.
> Let vernal airs thro' trembling osiers play
> And Albion's cliffs resound the rural lay.

This is truly pathetic fallacy country.

Figure 2.8 English rural landscape: North Yorkshire

Theorists, such as Addison, were adamant that the works of nature were best 'the more they resemble those of art. Hence it is that we take delight in a prospect which is well laid out, and diversified with fields and meadows; woods and rivers ... in anything that hath such a variety or regularity as may seem the effect of design' (Marx 1964: 93). Addison's 'improved' country-side, in its variety, was good prospect–refuge country, harked back to the designs of the Italian Idealists, and placed man firmly in the dominant role as improver and judge.

For a variety of reasons, which I will discuss in detail later when dealing with mountains and wilderness as landscape types, the semi-formal, semi-symmetrical classical notions of the beautiful were gradually overtaken by a growing feeling for dis-harmony, dis-proportion, and irregularity. These, all cardinal sins in the classical canon, were nevertheless derived from the idealist Salvator Rosa's wild classical landscapes (Figure 2.5). The Grand Tour, crossing the rugged Alps with difficulty, also favoured a gradual rejection of the smooth in favour of the rough.

Rosa's paintings dramatized mountains, deserts, waterfalls, crags, preci-pices and rough seas. Given the general feeling for the beautiful, these sublime landscapes were at first a minority taste. Gradually, however, the almost-medieval horror and fear of the sublime gave way to awe, and by the later eighteenth century both mountains and wilderness had come to be regarded as aesthetically pleasing. The concept of the sublime reached its zenith in the early nineteenth-century Romantic movement, in the poetry of Byron, and in the paintings of Turner. It is found today in our cult of wilderness.

A subset of the sublime was a growing taste for the Gothick (Figure 2.9). From Walpole's *Castle of Otranto* and Mary Shelley's *Frankenstein* to Bram Stoker's *Dracula*, an aesthetic appreciation of horror emerged. Eerie castles set on crags; bats, vampires, werewolves, and mad scientists; thunder, lightning, and dark of night; all these now-familiar props came together in the Gothic novel. This extreme, and now camp, variation of the sublime is relived in numberless teenage horror movies, the elegant novels of Mervyn Peake, and the crudities of Stephen King.

Although it can be argued that all landscapes are feminine and under masculine control (Tiffany and Adams 1985, Schaffer 1988, Porteous 1989), eighteenth-century aesthetic theorists such as Burke and Kant tended to see the beautiful and the sublime as gendered landscapes (Nead 1992). Whereas the beautiful landscape is feminine, and promotes contemplative pleasures, its sublime counterpart has masculine attributes, and excites arousal. Pictur-esque landscapes, however, sidestep this issue by focusing upon form rather than function.

The notion of the 'picturesque' was developed late in the eighteenth century. The basic concept was initiated by the Reverend William Gilpin, and championed by Sir Uvedale Price and Richard Payne Knight, and became sufficiently popular to be satirized in the novels of Thomas Love Peacock and Jane Austen's *Northanger Abbey*. The fundamental concept was that the

Figure 2.9 The Gothick: the ruined abbey at Whitby, Yorkshire, haunt of Bram Stoker's Dracula

landscape ought always to be seen as if it were a picture. The picture in question was the type painted by Claude or Salvator Rosa, for Poussin was rejected as being far too formal, smooth, and classical. At the other extreme, the sublime was rejected as being un-English, for nowhere in Britain were there landscapes vast enough to compare with the Alps. The picturesque, therefore, lay between the stupendous vastness of the sublime and the overly smooth, gentle tameness of the beautiful.

Theorists of the picturesque set up a number of criteria for judging a landscape or a painting. These included 'roughness', 'intricacy', 'sudden variation', 'abruptness', 'mystery', and 'surprise', many of which later became important variables in twentieth-century townscape planning, landscape assessment, and psychological experiment (see Chapters 3 and 5). Detailed handbooks on how to group park animals in a picturesque manner (groups of three are best) and how best to develop ruins (Figure 2.10), and arrange rocks and trees, were written. Gilpin (1786, I: 67–8) on ruins gives the flavour:

> It is not every man who can build a house, that can execute a ruin. To give the stone its mouldering appearance – to make the widening chink run naturally through all the joints – to mutilate the ornaments – to peel the facing from the internal structure – to show how correspondent parts have once united; though now the chasm runs wide between them – and to scatter heaps of ruin around with negligence and ease; are great efforts of art.

Figure 2.10 The pleasure of ruins: column near Castle Howard, Yorkshire

Out in the countryside, landscapes were judged on their resemblance to the picturesque ideal (Hussey 1929).

Indeed, in order to see landscapes properly, as if they were paintings, picturesque travellers carried a Claude glass. This tinned convex mirror performed two functions. First, it put a frame around the landscape, and second, it tinted the view to resemble the golden-brown monochrome of a Claude painting. (Compare modern tourists stopping at a 'scenic viewpoint', viewing it through tinted sunglasses, and then all taking the same photograph, often with a coloured lens filter!) According to Thomas West's (1778) popular guide to the English Lakes, the Claude glass distanced nearer objects, provided a regular perspective, and softened nature's often gaudy colours. One used it to compose a correct, finished picture.

'The picturesque style aimed to enhance the variety and intricacies of the landscape so that it would be visually striking' (Evans 1995: 27). The results of picturesque theorizing were picturesque painting (Figure 2.11), picturesque gardening, picturesque poetry, and picturesque travelling with picturesque guidebooks, many of them written by Gilpin. The picturesque is alive and well today in suburbia, at scenic viewpoints, on picture postcards, calendars, and chocolate boxes, in millions of holiday snaps, and in the endless cheap oil paintings (all of which seem to depict a deer by a stream among trees; real prospect–refuge stuff!) turned out on assembly lines in Hong Kong.

The Picturesque movement was remarkable in the development of environmental aesthetics for its concern with nature in the raw. For the first time, nature was looked upon as the equal of a painting. Landscapes were now admired for themselves, rather than (as in the beautiful and the sublime) for their ability to arouse emotions. The Picturesque movement drew people outside their gardens and parks to experience nature and the countryside, and exercised the sight, so that people began to form 'the habit of feeling through the eyes' (Rees 1975). Its most unfortunate result is its part in developing our modern tourist tastes, which disregard the complexities of nature in preferring a mere view of scenery. The early picturesque traveller, venturing out from his garden, had an involvement with the landscape which has been lost by modern tourists.

The Romantic movement

To the modern urbanite nature is a nuisance, a resource, or a picture (Rees 1975). Yet much of our schooling, in poetry at least, still depends on Romantic nature poetry, which had entirely different feelings about nature.

The movement developed in the late eighteenth century and flowered in the early nineteenth. It had roots in the picturesque, in the sublime, and in the development of science, for many of the Romantic poets and painters were well-versed in botany, zoology, geology, and meteorology. Philosophically, roots are to be found in the development of nature religion following the French Revolution, in Rousseau's concept of the noble savage, and in

Figure 2.11 The picturesque: M. Stanley's frontispiece to T. Lauder's 1842 edition of Sir Uvedale Price's *Essay on the Picturesque*

Goethe's botanizing, colour theorizing, and romanticized travel. A huge English and Scots Grand Tour phenomenon was by this time enveloping Southern Europe and beginning to penetrate Greece and the Levant. Among the best-known Romantic poets and painters are Byron, Wordsworth, Shelley, Keats, Coleridge, Constable, and Turner.

The chief tenets of Romanticism were idealist and religious. Nature manifests God. Indeed, nature is itself divine. A metaphysical quest was necessary to discover the divine essence manifested in natural forms. Further, nature was seen as exerting a 'moral force' on humankind. These notions have some similarity to the concepts of 'primitive' peoples but very little relationship to medieval ideas of nature as symbol. Equally unmedieval was the belief in the possibility of a spiritual union between humans and nature. Romantics enjoyed an emotional relationship with landscape, so that earlier notions of humankind vs. nature became the less planetarily dangerous concept of humankind-in-nature. Accordingly, Romantic Christianity took on pantheistic forms. Thoreau's 'I cease to live and begin to be' expresses this feeling of oneness with the natural landscape.

Romantics, too, were among the last to uphold the possibility of the marriage of art and science. Both, after all, used the same raw materials; together they could generate a greater understanding of reality. This belief, now unfortunately lost, coloured the sprawling worlds of Alexander von Humboldt, Goethe, and Ruskin. Equally salient to the Romantics was the importance of understanding the world experientially, as well as through the lenses of science. For them, remote sensing by satellite would be far inferior to intimate sensing on the ground (Porteous 1986). Above all, the Romantics were vitally concerned with learning to see. Their attitudes influenced poetry, novels, painting, and tourism far into the twentieth century, and thus it is fitting that we consider some Romantic exponents in detail, notably Constable, Turner, and Wordsworth.

Constable had a passion and an appetite for nature and an aptitude for scientific enquiry. His observations of the weather on the backs of his canvases are so detailed that they can be used to date the paintings (Thornes 1979). Constable was noted for his intense field observation and his endless experimentation to achieve the best match between painting and landscape. This is the more objective, empirical end of the Romantic spectrum. His paintings are soft, bucolic, romantic. Prime examples such as *The Hay Wain*, *The Cornfield*, and *Willows by the Stream* reconstruct ordinary English landscapes, mainly of the Stour valley in East Anglia. In his own words, Constable wished to capture 'the sound of water escaping from mill-dams, willows, old rotten planks, slimy posts and brickwork', just the ordinary landscapes so execrated by Michelangelo when viewing earlier Dutch examples.

While Constable derived his bucolic art from notions of the picturesque and the beautiful (Figure 2.12), Turner's work was sublimely wild (Figure 2.13). He had a strong sense of 'nature's unsubduable force', manifested in the depiction of mountains, avalanches, whirlwinds, stormy seas, waterfalls, and rough sky. There was no pretence of objectivity, yet Turner claimed he painted merely what he saw. The Turnerian view dominated late nineteenth-century landscape art, and even allowed the depiction of the machine in the landscape, with huge dramas of burning ships, rushing steam engines, and even the romantic, almost gothick horror of the flames and smoke issuing from industrial furnaces.

These two sides of Romantic painting coexisted. Constable's 'moral perfection of the familiar' depicted landscapes in which people would like to live, the rural idyll that lies behind the modern development of suburbia. In

Figure 2.12 The bucolic: John Constable (1776–1837) *Cottage in a Cornfield* (by courtesy of the Board of Trustees of the Victoria and Albert Museum, London)

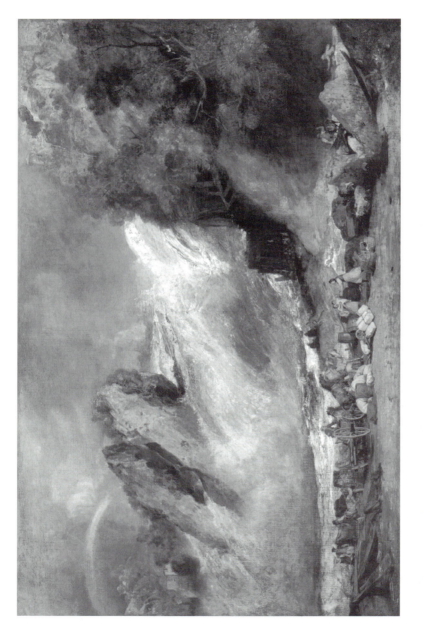

Figure 2.13 The wild: J.M.W. Turner (1775–1851) *Falls of the Rhine at Schaffhausen* (Bequest of Alice Marian Curtis and Special Picture Fund: courtesy, Museum of Fine Arts, Boston)

contrast, Turner's excitement and passion for the unfamiliar taps another strain. The Turnerian landscape, exotic and dangerous, is one we would like to visit, experience, and then leave. The Romantics had a strong feeling for the exotic, a feeling, sadly sanitized, which drives both the *National Geographic* (Lutz and Collins 1993) and tourism today. They also readily

Figure 2.14 Aesthetico-political Romanticism: Welsh bard curses English invaders. John Martin (1789–1854) *The Bard* (courtesy Tyne and Wear Museums, Laing Art Gallery, Newcastle upon Tyne, England). See also Thomas Gray's (1716–71) Pindaric ode *The Bard*

combined the aesthetic with the political (Figure 2.14), an issue taken up below in Chapters 4 and 5.

The nature-worship theme, however, is most explicitly discovered in poetry, and most notably among the Lake Poets (Wordsworth, Coleridge, *et al.*). Ordinary things were important to Wordsworth: stones, daffodils, a spade, a pair of boots. The Lake Poets learned humility and respect for the everyday, believing that the Whole was manifest in its smallest parts. Emotional contact with nature was essential; affection was required to understand nature authentically. Ideally, childhood training should awaken the child's emotions and guide her attachment to beautiful and worthy objects. Wordsworth gave no practical directions on how to achieve these states of being; one must go out for oneself and develop reverence for, humility towards, and a sense of harmony with, nature.

Wordsworth clearly felt that nature was a teacher:

> One impulse from a vernal wood
> Will tell you more of man;
> Of moral evil and of good,
> Than all the sages can.

He and his fellows were reacting strongly to the glorification of intellect which prevailed during the eighteenth-century Age of Reason. They preferred to inhabit, like Emerson and Thoreau, a deistic universe where woods, trees, rocks, and clouds became scripture.

In *Lines composed a few miles above Tintern Abbey* (1798) Wordsworth describes the green, pastoral landscape, suggests that this view of nature may predispose people to good actions, and adverts his inner state, a sense of oneness with the divinity of nature:

> I have felt
> A presence that disturbs me with the joy
> Of elevated thoughts; a sense sublime
> Of something far more deeply interfused
> Whose dwelling is the light of setting suns,
> And the round ocean and the living air,
> And the blue sky, and in the mind of man:
> A motion and a spirit, that impels
> All thinking things, all objects of all thought,
> And rolls through all things.

Here Wordsworth is experiencing an epiphany, an oceanic feeling, Maslow's (1954) 'peak experience', a feeling of deep, complete correspondence between inner state and outer landscape which Seamon (1976) has described as 'spiritual ecology' (see Chapter 3.4).

Modern times

Among the elite, the notion of the sublime had died by the end of the eighteenth century, the picturesque by the early nineteenth century, and the Romantic after the mid-century. Romanticism was killed by the modern over-development of science and technology. Specifically, the notion of a divine, moral nature could not survive Darwin's *Origin of Species* (1859). Further, the invention of the camera in 1826, and its gradual dissemination to the masses, made landscape painting redundant and allowed the many to indulge their tastes for the sublime and the picturesque, as these ideas, abandoned by the elite, trickled down the cultural gradient. And most finally, the triumph of science and technology and their scorn for mystical thinking of any kind cut off the spirit of the age from both art and religion. As Western societies developed the tools for making over the earth, the 'feel' for harmony between humankind and nature was lost in the frantic exploitation of 'resources'.

At this point also the widening gap between ordinary and elite taste began to make itself felt. According to Clark (1956: 142), 'never before has there been such a complete divorce between popular and informed taste'. The latter has made its way through Impressionism ('the painting of sensation', 'psychologists working with paint') and Post-impressionism, where all pretence of naturalism was abandoned in favour of the idiosyncratic vision. By this time we had lost any notions of religious or natural order and were ready for the Modernist early twentieth century, with its mechanic cubists, Surrealist dreamscapes (Figure 2.15), line and geometry Expressionism, Wasteland poetry, the music of Stockhausen, and the Theatre of the Absurd. Or, perhaps, the fragmented late twentieth-century Post-It world of post-industrial, post-religious, post-communist, post-capitalist, post-Modernism for which, apparently, 'the sublime is now a buzzword' (Nead 1992: 29). Things fall apart. But not for the general public, who are still enjoying the trickle down of eighteenth and nineteenth-century elite notions as, in picturesque and Romantic modes, they decorate their interiors, design their gardens, and take their holidays.

Figure 2.15 A (real) surreal landscape: radar station on Fylingdales Moor, Yorkshire, gives four minutes warning in the event of nuclear war

2.3 LANDSCAPES

Having looked at changes in landscape tastes in terms of historical periods, it is now time to take a more detailed view through a discussion of the changing appreciation of major landscape types. These landscape types are: mountains; wilderness; the middle landscape (rural); gardens; and townscape.

Mountains

Nicholson's *Mountain Gloom and Mountain Glory* (1959) takes its title from John Ruskin, who had earlier (1886) divided ancient from post-seventeenth-century views of mountains thus. Although it is now normal to enjoy looking at mountains, this has not always been so.

Classical, medieval, and early Renaissance painters and poets either ignored mountains or had an active hatred for them. Yet their journeys caused them to frequently cross the Alps, whose mountains, chasms, ravines, crags and torrents were seen as 'places of torment', an earthly hell. Travel through mountains was dangerous, and no one climbed them for pleasure. We have already seen what happened to the poet Petrarch when he aberrantly scaled Mount Ventoux in 1336. Mountains were seen as the homes of hobgoblins, spirits, and dangerous beasts, a feeling which persisted even in Switzerland until the sixteenth century.

By the seventeenth century this fear and horror of mountains had been explicated in religious terms. Thomas Burnet, in his *Sacred Theory of the Earth* (1684) contended that:

> The Face of the Earth before the Deluge was smooth, regular, and uniform, without Mountains, and without a Sea ... This Smooth Earth ... had the beauty of Youth and blooming Nature, fresh and beautiful, and not a Wrinkle, Scar or Fracture in all its Body; no Rocks or Mountains ... but even and uniform all over.

Change came, of course, because of human sin, later washed away by the Flood, which left the face of the earth all warts, pockmarks, wrinkles, and age-spots, the 'Ruins of a broken World'.

Here theology was married to Classical aesthetics. The earth had once had smoothness, regularity, uniformity, proportion, roundness, and symmetry in full measure. Sin had resulted in its becoming rough, irregular, craggy, asymmetrical, diverse, and various, all of which were anathema to the classically-trained mind (though later the inspiration of the Romantics).

The change to a positive view of mountains came about through science and new religious ideologies. With the invention of both telescope and microscope humankind became readier to accept the nature of things without religious preconceptions. People also became aware of vastness. Practical arguments emerged, noting, for example, that mountains create useful harbours and generate clouds and rain. Further, the new theology of the late seventeenth century argued that the existing earth was God's doing, and therefore that variety and diversity were good in themselves. It was also considered to be presumptuous to apply the theories of classical aesthetics to God's work, as if the theologian were aware of God's aesthetic criteria.

The general result of this reappraisal was that the object of the human capacity for awe (defined as a mixture of terror and exultation) passed from God to the cosmos (via the telescope) and thence to the greatest objects in the geocosm, notably mountains, oceans, and desert. This led to a new feeling on the part of those crossing the Alps on the Grand Tour. Absolute terror became 'an agreeable kind of horror' (Addison), a 'delightful horror', or 'a terrible joy'. There was still the same danger, given no great improvement of horse-and-carriage technology, but the danger was now looked upon with some satisfaction. Mountains, then, with their vast, rude, misshapen bulks, became sublime and magnificent with the general change of landscape consciousness in the eighteenth century.

The idea of Alpine sublimity was soon transferred to Britain. Wales, Scotland, and Northern England, shunned by travellers in the early eighteenth century, quickly became travel magnets. By mid-century sublime mountains had become equal in attraction to beautiful lowlands, and were deemed far superior by 1800. Wordsworth's *Guide to the Lakes* (1835) noted that: 'A stranger to mountain scenery naturally on his first arrival looks out for sublimity in every aspect that admits of it.' By the Romantic era, then, the sublime experience of mountains had become self-conscious, anticipated, and touristy.

The high intensity of feeling associated with the sublime aspect of mountains was significantly reduced in Britain with the rise of the late eighteenth-century Picturesque movement, which denied the existence of the sublime in Britain. Later, mid-nineteenth-century transportation improvements enabled the conquering of mountains, deserts, and seas, thus reducing their sublimity to a lower-intensity beauty or picturesqueness.

Wilderness

Whereas the sublimity of mountains was a European concept, the positive notion of wilderness is more clearly North American (Nash 1973). Wilderness is a qualitative term which involves wild, uncultivated, 'unspoiled' land inhabited by wild creatures, and where humans are merely visitors. Early concepts of wilderness derive from the Biblical notion of a desert place, a negative area, the abode of demons, as well as from the medieval Teutonic fear of dark, primeval forests.

When the Puritans began to settle New England in the seventeenth century their major aim, as frontier people, was to conquer the wilderness and turn it into a pastoral garden. The North American forests were seen as the unknown, as evil, beast-ridden, wild counterparts of the former forests of a Europe now tamed. When William Bradford stepped off the *Mayflower* he found 'a hideous and desolate wilderness', the enemy of civilized life, which it was his duty to conquer, subdue, and vanquish. Wilderness was seen also as a sinister moral vacuum, a cursed wasteland full of heathen to be tamed and Christianized.

A major fear was that the freedom provided by the wilderness would soon strip off the veneer of civilization, giving humankind the opportunity to once more become savage and bestial. Shepard (1967), echoing Nathaniel Hawthorne's *The Scarlet Letter*, sees the Puritans' fear of wilderness as a reflection of their disdain for sexuality. Wilderness, like sex, was difficult to control, frenzied, and not easily amenable to rules and regulations. The Puritans came looking for a second Eden, but were confronted, just beyond their pale of settlement, by a wilderness which was not only an actual threat to survival but also a dark, sinister symbol of evil.

A change in landscape consciousness came about in the United States in the early nineteenth century, chiefly as a result of the importation of European ideas into eastern cities. These were the first to absorb Rousseau's idea of the noble savage, the concept of the sublime, the decline of Judaeo-Christian fundamentalism, and notions of the corrupting influence of urban life. In contrast, wilderness, like rural areas, came to be seen as far more wholesome than city life, while Romantic deism believed that God's power and excellence were most obviously manifest in wild places. As early as 1803 a gentleman explorer in the Ohio valley could say: 'There is something that impresses the mind with awe in the shade and silence of these vast forests. In the deep solitude, alone with nature, we converse with God' (Nash 1973: 58). Thus the notion of wilderness as a positive aesthetic environment flowed westward from the eastern cities, never quite catching up with the trans-Appalachian pioneers who still stoutly maintained that the wilderness was 'howling', 'dismal', or 'terrible', and should be remade so as to 'blossom like the rose'.

Nevertheless, this 'hatchet view' of wilderness eventually succumbed to the flood of Romantic paintings, poetry, and travel books which gushed from the tamed eastern seaboard. Explorers such as Estwick Evans (1818) deliberately made tours in winter, in order to 'experience the pleasure of suffering, and the novelty of danger'. 'Meditation on nature [became] a recognized avenue to the centre of being .... "Looking" became an act of devotion' (Novak 1980). Thus we find Thaddeus Harris perched on a mountain crag in 1803, ambivalently looking east towards 'fruitful fields' emerging from 'dreary forests', yet also looking west towards the untamed forest, where 'the sublime in nature captivates while it awes, and charms while it elevates and expands the soul'. In these schizophrenic terms the westward-moving American frontier can be seen to divide a tamed eastern paradise, the beautiful landscape of Puritan imagination, from an untamed wilderness regarded as positive (sublime) by the aesthetic elite and negatively by the working pioneer.

Significantly, the elite were conscious of the fact that wilderness was uniquely North American, having all but disappeared from Europe. In the latter, the glories of a Thoreauvian life in the woods were no longer possible.

Conscious of their cultural inferiority, American elites hymned the virtues of the simple, moral life and its landscape counterpart, the wilderness. It was not long before American boasting of the size and splendour of their landscapes' sublimity became a commonplace, Europe being reduced to the merely picturesque. America perhaps had little history, but here was a continent 'fresh from the hands of the Creator', a holy place, 'the unedited manuscript of God'.

Just as F.J. Turner believed that the frontier experience served to mould American character, so it was believed that the experience of wilderness would render Americans more innocent and moral than Europeans. The wilderness, after all, was wild, noble, innocent, moral, aesthetic, vast, and sublime, attributes to be preferred in a powerful expanding nation. This spiritual union of landscape aesthetic and national character was a major force behind the national parks movement. As early as 1832 George Catlin, a major American landscape painter, was calling for 'a Nation's Park ... in all the wildness and freshness of ... nature's beauty ... to preserve and hold up to the view of her refined citizens and the world, in future ages' (Nash 1973: 101).

The ideas of John Muir, Frederick Law Olmstead, the National Parks movement, the Sierra Club, and the Wilderness Act (1964) are later examples of this deep-seated ideology of wilderness purity. Graber's (1976) study of modern wilderness purists demonstrates that many aspects of their faith are quite comparable with the sublime and Romantic notions of almost two hundred years before. The purist gains an almost religious experience, 'a sudden illumination of individual consciousness', from personal contact with the 'sacred power' of wilderness. Wilderness is seen as an earthly version of sacred order and perfection, where humankind is best able to achieve transcendence. These ideas are variously displayed in the writings of Thoreau, Joseph Wood Krutch, Aldo Leopold, and Edward Abbey, and the photography of Ansel Adams and Eliot Porter, and have been effectively parlayed in the political arena by activists intent on a greater degree of preservation of the remaining American wilderness (Chapter 4). Elsewhere, intense efforts are being made to re-create wilderness landscapes of forest or prairie, although such reconstitutions are usually branded as fake by wilderness purists (Turner 1988).

The middle landscape

Between mountain and wilderness on the one hand, and garden and city on the other, lies the middle landscape of agrarian enterprise. This approximates to the pastoral or bucolic ideal, preferred by eighteenth-century idealizers of the beautiful, and earnestly worked for by tree-hewing Puritans.

Classical poets almost universally celebrated a wholly cultivated, pastoral, tamed nature, where the beautiful was strongly related to the useful. The poet Lucretius (first century BC) contrasted rural beauty with the great natural 'defects' of the mountainous and forested wilderness from which the agrarian landscape had been wrested. For these thinkers, only human inventions, such as orderly cultivation, enabled humans to rise above, or escape from, the perils of the wilderness. Greek and Roman poetry is full of pro-agricultural epics. Horace and Virgil already felt the need to flee the pestilent cities for a more 'natural' way of life, not in the forests, but among arable and pastoral, crops and sheep, yeomen and shepherds. Lucretius was glad to see wilderness retreat in favour of 'fields, crops, and joyous vineyards, and a grey-green strip of olives to run in between and mark divisions ... adorned and interspersed with pleasant fruits, and fenced by planting them all around with fruitful trees'. This is Arcadia, the happy mean between city and wilderness, a fruitful, pastoral idyll where sheep may safely graze.

Preference for the middle landscape also had deep Biblical roots. The Old Testament is rife with negative references to wilderness (unless as a place for prophets to meet God), positive references to sheep-rearing, and the joy of raising crops. 'The Lord is my shepherd' is matched by a psalm in which 'The valleys also shall stand so thick with corn that they shall laugh and sing.' When Jehovah chose to chastise the Hebrews he threatened to lay waste the hills, and dry up the herbage. When he wished to confer blessings, he promised to transform the wilderness into a land of brooks and water, fountains and springs. In Isaiah especially we find prophecies of the coming of 'streams in the desert' so that 'the desert shall rejoice and blossom as the rose'. Israeli technology, of course, has helped bring this to pass.

Other historic strains which lie behind this great preference for the middle landscape include the delight in gardens (to be dealt with later) and the folk history of medieval Europe, dominated by the great epic story of the destruction of primeval forest in favour of productive arable. It is no accident that the hero of fairy tales is usually a woodcutter.

Small wonder, then, that rural environments have been idealized in Europe from classical times until the twentieth century. It is natural also that this deeply ingrained tendency should be transferred bodily to North America with the early settlers, whose fears of the wilderness were balanced by their joy in being able to transform it into fruitful fields. From the late eighteenth century until the middle of the nineteenth, the preferred American landscape was rural, the Jeffersonian ideal. Jefferson wanted a nation of innocent, happy, bucolic yeoman farmers with little industry and only small towns; his ideal was very classical in tone. Here, then, we have the typical New England tourist-poster village, an ideal later to be transported as far afield as Hawaii and Tonga.

The middle landscape's problem lay in its middle position. As agriculture chewed away at wilderness, so the cities sprawled over the countryside (Figure 2.16). Here we find Marx's (1964) evocative *Machine in the Garden*, the invasion of the pure, the sacred, the pastoral idyll by noisy, filthy, crude, mechanical, inorganic, commercial, crass industrialism, as cities spread and industries moved out to rural sources of water power (Figure 2.17).

This American anti-machine feeling, like the pro-rural and pro-wilderness feelings before it, was imported from Britain. As early as the beginnings of

urban source of rural
and wilderness idealism ⟶

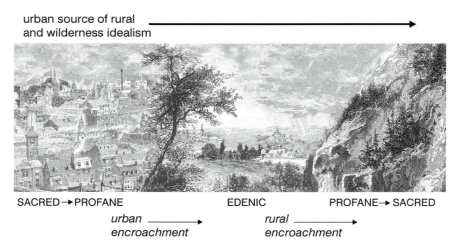

SACRED→PROFANE   EDENIC   PROFANE→ SACRED

urban ⟶   *rural* ⟶
encroachment   *encroachment*

Figure 2.16 American landscapes in the nineteenth century

the Industrial Revolution William Blake had castigated the 'dark satanic mills' of industry. But the anti-industrial aesthetic did not come into sharp focus until the middle of the nineteenth century. Marx draws attention to two literary incidents of 1844 which express the elite's belief that the inorganic was submerging the organic. Wordsworth, opposing the projection of a railway through the Lake District, cries: 'Is there no nook of English ground secure/From rash assault?' And Nathaniel Hawthorne, his quiet shattered by

Figure 2.17 Modern industry invades the countryside: cooling towers in Yorkshire

a New England steam-engine whistle, deplores this bringing of 'the noisy world into our slumberous peace'. Both feared that this 'fever of the world', a world wholly devoted to 'getting and spending', would, in the name of 'progress', promote a fragmented, inorganic, industrial lifestyle at the expense of the beautiful, healthy, agrarian landscape. They were quite right.

During the period 1840–60, a period designated by the economist Rostow (1960) as America's 'take-off into sustained growth', a whole genre of writing on the theme of pastoral destruction by industry emerged. Rural life, honest, strong, upright, productive, regenerating, clean, moral (all, incidentally, virtues acclaimed by groups as various as right-wing conservatives and the 1960s back-to-the-land movement), 'a place apart, secluded from the world – a peaceful, lovely, classless, bountiful pasture – a well-ordered green garden of continental size' (Marx 1964), was being subverted by the oppression of the European factory system.

By the mid-nineteenth century, parity between the rural and the machine was not enough, and the machine celebrants took the offensive. Machine civilization was touted as the only way to progress, to tame the wilderness, to civilize a continent. The great myth of the civilizing influence of the railroads was developing, with George Inness in 1854 painting scenes for the Lackawanna railroad, forerunner of later paintings of American landscapes that are totally inorganic, a celebration of silos, cranes, and factories.

A new pastoralism had, in fact, emerged, an industrialized pastoralism which saw the machine as an ally in destroying wilderness to create more middle landscape. Poems about 'iron men' with 'heroic wills' were matched by railroad folk-songs, all celebrating the power of the machine over brute nature. Hence the machine was able to co-opt the pastoral ideal and transform it into a celebration of inorganic power. This cult of the machine entered the twentieth century with the poetry of Sandburg, the paintings of the Futurists, and the architecture of Le Corbusier. In its 'high-tech' computer-driven form, it is still with us.

The paradox of the machine cult is that the general public, though now almost all living in urban regions, have a strong predilection for the rural idyll. In Britain, there is a strong mass base for the defence of 'the countryside' against obvious encroachment, the North American counterpart being wilderness preservation and the glorification of small-town life. In both regions, there was a great rush to the 'pastoral' suburbs after World War II, and after 1970 a strong counter-urbanization trend in favour of villages and small towns. The ideal American environment, according to polls, is still that of the small town set in a rural landscape.

Landscape gardening

Landscape gardening forms a bridge between, on the one hand, genuine nature or nature transformed for economic purposes, and on the other, works of art such as paintings, poetry, and even cityscapes. The landscape garden is nature transformed into an idealized conception of landscape form.

Its relationship to the world beyond the garden has changed with alterations in our feelings about that external world.

In ancient and medieval times, the garden was seen as 'a paradise', the word paradise being derived from the Persian for 'walled enclosure'. Indeed, early gardens were walled off from the outside world, whether from the Egyptian desert or the forbidding mountains surrounding the cloisters of medieval monasteries. The first description of an English medieval garden, *c*.1250, involves high walls, hence much privacy, a fishpond, a herbiary, and grassy walks. It is as much economic as recreational in tone. Gardening books had appeared by 1440, and intricate geometrical garden forms, such as mazes, labyrinths, and knot-gardens, had been developed (Figure 2.2).

This emphasis on the artificial, nature cut to human size and pliable to human whim, continued in the Renaissance pleasure garden (*c*.1660–1725). Here the greatest influence was Le Nôtre, royal gardener to Louis XIV, who began Versailles in 1661. Charles II wished to borrow Le Nôtre from Louis, and the 'French Style' became all the rage in late seventeenth-century Britain. These gardens were characterized by their uniformity and rectangularity, with endless vistas punctuated by statuary. Nature was severely regimented, with hedges trimmed to shape and much topiary work. A common plan was the goose-foot, where gravelled avenues or canals, lined by hedges, fanned out from a central Baroque palace. These Baroque vista gardens were regarded as outdoor rooms, and this idea was sometimes extended into nearby forests by cutting long straight swathes, or 'rides', through them. Flowers were banished in favour of clipped evergreens, notably the dark holly, box, and yew. The result was endless flatness – grass lawns, gravelled parterres, clipped hedges, long avenues and canals, as if the whole garden had been transformed into a kind of outdoor ballroom (Figure 2.18).

By the end of the seventeenth century most of these geometrical gardens were still surrounded by high walls, beyond which lay unkempt countryside, much of it waste (Stroud 1954). Two transformations occurred in early eighteenth-century England. First, the enclosure movement transformed the vast open fields into the small hedged fields so typical of modern British tourist posters. Second, the great landowners tired of the tedious geometry of their formal gardens and began to look elsewhere for their gardening models.

The Grand Tour introduced them to the idealized landscapes of the Italian school, of which the classical Arcadias of Claude were found most impressive. The idea of transforming their geometrical gardens into Claudeian Arcadias appealed immensely to a generation imbued with classical art and language. The chief theorist of this movement was Horace Walpole who, as fourth son of the British Prime Minister, Sir Robert Walpole, had considerable influence. In his *History of the Modern Taste in Gardening* (1788), Walpole first denigrates the Baroque style of Le Nôtre and then extols the new, 'naturalistic' style. His work is worth looking at in some detail.

Versailles, thought Walpole, was 'a garden for a great child', and the geometrical designs brought over to England from Holland by King William

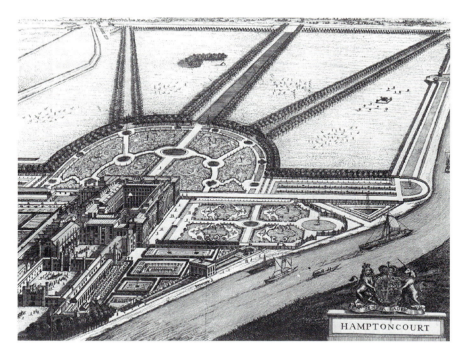

Figure 2.18 The Baroque landscape at Hampton Court, near London, *c.*1690

were little better. Walpole traces the development of walled gardens from the Classical period, and notes that they universally exclude 'nature and prospect', that symmetry is the god (Pope's 'each alley has a brother/And half the garden just reflects the other'), and that the whole effect was of 'the same tiresome and returning uniformity'. He then considers the parks lying outside the garden walls, their more natural appearance being deemed more pleasant to the eye. Unfortunately, England's low topography, unlike that of Italy, afforded little in the way of vistas over these garden walls.

The need then, was to develop pleasing irregularity in the deliberate creation, in concrete terms, of Milton's 'happy rural seat of various view'. The most powerful initial stroke, invented by the gardener Bridgeman (*fl. c.*1730) was the destruction of the garden walls, hallowed by several thousands years of use, in favour of hidden-trench boundaries known as fosses or ha-has (so-called on account of one's exclamation on suddenly encountering such a ditch). This 'sunken fence' had the admirable function of separating park and garden, thus keeping animals from the house, yet at the same time allowing an observer at the house itself to gaze uninterruptedly and dominatingly across a mighty vista consisting of garden, park, and rural landscape beyond (Figure 2.19).

William Kent (d.1748) made the next move. According to Walpole, 'He leaped the fence and saw all nature was a garden.' Under Kent's tutelage, all Baroque extravagances were swept away, and the garden became

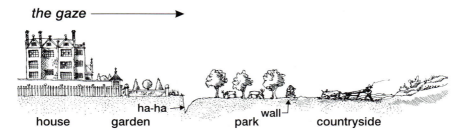

the gaze ⟶

house   garden   ha-ha→   park   wall⏋   countryside

Figure 2.19 The uninterrupted gaze from country house through garden and park to rural landscape beyond

indistinguishable from the park, a vast rolling lawn broken here and there by clumps of trees. The general aim was to realize, on the ground, artful compositions equal to those of the greatest Italian landscape painters. Nature was altered subtly, rather than abruptly in the Baroque manner; 'The living landscape was chastened and polished, not transformed.' Kent's ideas were carried to greater extremes by Capability Brown (d.1783) and Humphrey Repton (d.1818). Their general effect was to create a wholesale transformation of English gardens, estates, and much of the countryside visible from the great country houses (Figure 2.20). The classically beautiful sweeping lawns of Brown (Figure 2.21) and the picturesque asymmetry of Repton were regarded by Walpole as an ideal English landscape and a model for the world. Indeed, the former has become a major tourist attraction and the model, copied in small throughout the Western world, has been followed endlessly in the vast lawned suburbs of modern cities.

Eighteenth-century landscape gardens were built upon three major principles. First, 'the concealment of bounds' meant that, by using the ha-ha, the eye could run from house, across garden, park and countryside, uninterruptedly to the horizon. An immense feeling of openness, and of power, was derived from this vista. Second, 'the management of surprises' involved unexpected turns, interrupted vistas, and curves along walks in other parts of the garden, thus generating a pleasant feeling of mystery. Third, 'the provision of contrasts' ensured that the 'natural' should be offset by the artificial in the form of classical temples and the like.

The chief elements of such a garden would include water, not rectangularly laid out in canals, but in serpentine streams and irregular pools, classical temples at vantage points, rather than at compass points (Figure 2.22), and gothick ruins, often especially built, where the stroller could contemplate grandeur overrun by barbarity, or religion destroyed by kingly power. Especially effective were cases, as with Fountains Abbey – Studley Royal in Yorkshire, where a landowner was able to include a genuine monastic ruin in his landscape.

'Borrowed landscape', so frequently used in Japanese gardens, was also involved. Here views within the garden were oriented towards objects, such as church spires, beyond the estate's bounds. Extrinsic elements such as

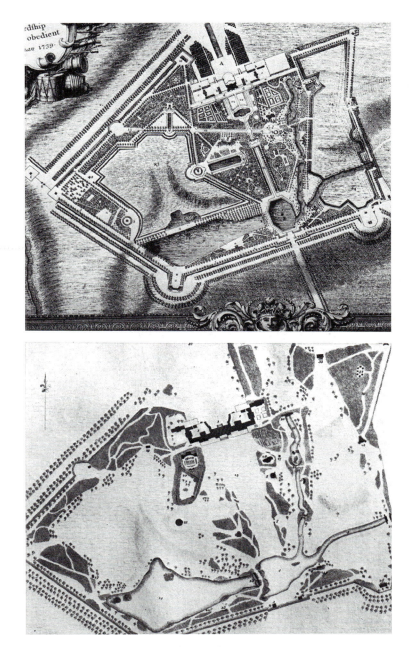

Figure 2.20 The transformation of gardens and park at Stowe, Buckinghamshire.
Above (c.1739), formal design by Vanbrugh and Bridgeman. Below (c.1797), as
redesigned by Kent and Brown

Figure 2.21 Heveningham Hall, landscaped by Capability Brown in 1781

Figure 2.22 Classical pavilion in Fountains Abbey gardens, Yorkshire

churches, monasteries, and attractive forms would be included wherever possible, whereas other extrinsic elements, such as whole villages, would be destroyed in order to clean up the view. Mood-management, other than surprises, was involved in the construction of grottoes in which melancholy 'hermits' were paid to reside.

Stroud (1954) notes that all this must not be taken overly seriously. In an age with no mechanical pastimes, taking the circuit walk through a landscape garden was an important source of entertainment. One walked or drove through, showed the garden to friends, remarked on their reactions, had tea in Greek summer-houses, and took with a pinch of salt the varied reactions to melancholy hermits, surprise views, and the 'horrid', 'gloomy' wilder parts of the estate. A kind of outdoor theatre was in operation, and for the first-time visitor at least a form of mood-management would be quite effective – tranquillity along the lawn by the smooth river, agitation in the darker woods or seemingly-unkempt 'wilderness', surprise on rounding a corner to see a ruin, delight at a view of a temple-crowned crag, terror at attaining the summit of that crag, and reflection at the sight of a far-off, fortunately unsmellable, ploughman slowly toiling through the fields. It is in this form of landscape gardening that all the arts of the eighteenth century come into contact – landscape theorists, painters, poets, politicians, architects, novelists, all these had some hand in shaping this major aesthetic landscape form. And economists too, for Moggridge (1986) has demonstrated that some of the designed facilities in parks were concerned less with aesthetics than with productivity.

By the mid-nineteenth century the great Age of Landscape was long over.

The Industrial Revolution, the breakup of great estates, a lessened interest in planting for generations to come, less leisure, and a greater orientation towards cash rather than land as a source of power, all these stifled any large-scale evolution of the eighteenth-century landscape garden. As a result, much of its ethos is still with us in our suburban layout of detached houses surrounded by lawns, with irregular trees and shrub borders, and in the turf-and-tree urban parks which are ubiquitous in the Western world.

Nineteenth-century gardening was influenced by the Picturesque movement, which called for greater irregularity (Figure 2.23), the invention of the lawnmower in 1830 ('amusing, healthy, and useful exercise' for country gentlemen), and the bringing back of flowers in great profusion. The latter was the nineteenth century's chief contribution. Gardens became 'perfectly dazzling' with hundreds of varieties of exotic flowers, most notably azaleas and rhododendrons, with orchids in conservatories. Trickle-down of tastes ensured that by the late nineteenth century both the lawn and flower beds had become an important part of the capitalist consumer system, and with the 'Wild Revolution' of that era a taste for English country gardens also set in. In the twentieth century, of course, eclecticism rules, with gardens ranging from the austere Bauhaus style to the desert xeriscaping of Arizona.

Townscape

Compared with our ready appreciation of natural and contrived rural or garden landscapes, we have much greater ambivalence about townscape, the landscape of cities. On the one hand, cities have long been regarded as humankind's greatest work of art (Mumford 1966). The orthogonal layout of early cities, their circular mandala outlines, and their central, heaven-reaching, man-made mountains in the form of temples, demonstrate a symbolic value that relates the urban handiwork of humans to the perpetual order of the heavens. On the other hand, modernist urbanism has often exalted the artificiality of the city. The extreme is Le Corbusier's dictum: 'The city is the subjugation of nature by man; it is a human action against nature' (Jackson 1959: 9).

Cities are distant from nature (Tuan 1978). Jackson (1959: 10) asks for a built environment which is as natural as possible, yet this becomes increasingly impossible when we build at so vast a scale that there is no longer any tradition to guide us. Many urban utopianist fantasies banish nature altogether, and the internalized hanging gardens of Portman-style hotels are no substitute for the real thing. City design and zoning policies have three major goals. In addition to social and economic efficiency and biological health, the city should provide its citizens with a 'continuously satisfying esthetic and sensory experience' (Jackson 1959: 11). Yet the outcries of urban critics, briefly touched upon earlier in the work of Smith (1977) and to be explored in detail later in the Activist section (Chapter 4), suggest that modern cities increasingly deny their citizens' right to sensory joy.

Figure 2.23 Landscape transformation from the smoothness of Brown (above) to the irregularity of the picturesque (below). From Richard Payne Knight's *The Landscape*, 1794

Urban environmental aesthetics has a ten-thousand year history which has been surveyed extensively elsewhere (Mumford 1938, 1966, Rasmussen 1951, Moholy-Nagy 1968, Bell and Bell 1969, Bacon 1976, Girouard 1985, Middleton 1987). What follows is an extremely brief synopsis.

The ancient city, whether Chinese, Egyptian, or Mayan, tended to be regarded as a cosmic umbilicus, an earthly point of access to heaven. Such early cities were chiefly known for their ceremonial rather than economic functions (Wheatley 1971, Smith and Reynolds 1987). Founding a city was a sacred act. The commonly-used grid street system symbolized conscious order, precision, and the imposition of regularity from on high via the local god-king. The centre of the city was the most sacred area, and here were erected the grandiose centralizing monuments of an all-powerful, heaven-sanctioned, earthly authority. Centralism and monumentalism, of course, were pervasive themes throughout the Classical period and re-emerged strongly in Western Europe during the Baroque period and in Nazi and Fascist states during the twentieth century. Centralized monuments are popular today in downtown office cores and in the capitals of newly-independent Third World states.

In the medieval era, monumental centralism was still represented as much by the church as by the castle (Figure 2.24). The typical medieval urban layout, however, had much more in common with contemporary Arab cities than with the formal grid so common in previous eras and spread across three continents by the power of Rome. Around the central spires and towers lay a seemingly haphazard network of narrow, winding streets lined by narrow-fronted houses and opening up only to form irregularly-shaped market-places. Unless kept open by market crosses, fountains, or regularly-held markets, open spaces would be built upon in a process of seemingly organic growth. Within the retaining wall, and best seen today in cities such as York, the meandering streets provided much visual variety and mystery via broken vistas and apparently unattainable landmarks (Figure 3.3). A much later century would see this layout as picturesque, and its organicist pattern would be repeated in the twentieth century in a host of middle-class suburban layouts replete with curvilinear streets and culs-de-sac.

In general, medieval cities were built as thousands of unrelated units, and achieved their aesthetic charm by accident rather than design. With the Renaissance, however, rather rigid principles of classical design were revived (Bacon 1976). The resulting cities, from fifteenth-century Italian utopias to seventeenth-century Baroque realities in northern as well as southern Europe, were extremely regular and formal in tone.

Some cities were almost completely rebuilt, not as a mass of unrelated units as before, but as deliberate compositions of buildings and related open spaces which together formed what were, in effect, tableaux. To comprehend the medieval street, with its sudden turns, half-blocked views, and general curvilinearity, the observer had to walk through it. This became unnecessary in the Baroque city, with its 'compositions in depth' involving a series of long, wide, straight axes all leading up to a central palace or cathedral.

Stationed at this central point, the ruler was able to comprehend the whole design in a single arcing glance. The planned Baroque city, designed as an aesthetic experience, was also a landscape of power par excellence.

Whole cities, such as Karlsruhe, were planned in this way, as were the visible countrysides around aristocratic country mansions. Methodical, regular, and geometrical, the typical design concentrated less on function than on architectural and decorative effect. Both city and its environs became theatre.

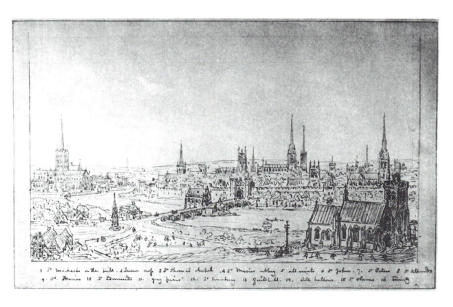

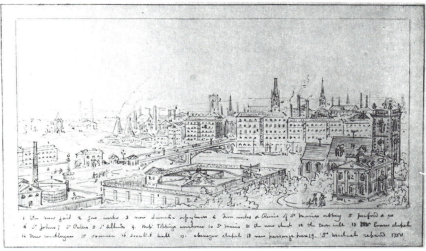

Figure 2.24 The city, dominated first by spires (1440) and then by ungodly factory chimneys (1840). From Pugin's *Contrasts*, 1836

These ideas were introduced to Britain by Inigo Jones in the early seventeenth century. The Baroque style's uniformity and symmetry, with large, regular sash-windows, *portières*, pediments, pilasters, columns and balustrades, was all very new to a population whose yardsticks were the irregular Tudor and Elizabethan styles. The effect of the intrusion can be seen in Sir Christopher Wren's design for the rebuilding of London after the Great Fire of 1666. His original plan, replete with huge palaces, vistas, and *rondpoints*, all in a uniform architectural style, was nothing less than an attempt to redesign the whole of London in the shape of a gigantic formal landscape garden (Figure 2.25).

Less ostentatious classical notions of uniformity came to the fore in northern Europe in the eighteenth century. Just as English rural landscapes and gardens were radically altered in the eighteenth century to fit Italian idealist models, themselves rooted in Greece and Rome, so in the same era the larger British cities were extensively rebuilt in the fashion of the seventeenth-century Italian architect Palladio, who had drawn extensively upon the Roman Vitruvius.

Palladianism, most popular in Britain in the period 1720–50, was a revolt against the gaudy Baroque which had recently erected on English soil the vast castellate piles of Blenheim and Castle Howard. Like the contemporary English landscape garden, the mid-eighteenth-century city exhibited a classic formalism which rejected Baroque ornamentation in favour of restrained and elegant forms. Long uniform terraces, a continuous eavesline, a regular fenestration of tall sash-windows, a very restrained use of pediments above doors, roofs partly concealed behind balustrades, wrought-iron gates and railings, the use of plaster or light-coloured paint so that all brick and stonework was covered, all these attributes were combined to produce the well-known Georgian façade with its characteristic unity, rhythm, harmony, and balance (Figure 2.26). Whole cities or city districts were built or reshaped in this manner, as the squares, crescents, circuses and serpentines of central London, Edinburgh, and Bath bear witness to this day. Contemporary landscape-gardening techniques were applied to fit contours; T-junctions were preferred to crossroads; houses were set back from roads to permit lawns and flower beds; and large patches of parkland began to appear. These characteristics, of course, remain common features of twentieth-century cities.

The immense growth of industry and population in the nineteenth century vastly altered this elegant, controlled landscape. The imposition of bylaws to set minimum standards for workers' housing meant, in general, that all subsequent housing was built to these minimal standards. And with the mass manufacture of bricks came total uniformity not only in layout and architectural style, but also in building materials. Victorian cities, whether in Europe or North America, tended towards the dull and boring.

Except for public buildings, for here the Victorians shone (Figure 2.27). The Victorian era, casting off all notions of a dominant style, embraced all known styles in an overlapping series of architectural revivals. Egyptian solidity for banks, Gothic authority for town halls, churches, and museums,

Figure 2.25 Medieval and Baroque city plans. Above, York; below, Wren's plan for rebuilding London, 1666

Figure 2.26 Georgian elegance, Regent Street, London. From J. Elmes's *Metropolitan Improvements*, 1829

Figure 2.27 Victorian exuberance, York, England

neo-Tudor for London's Houses of Parliament, classical Greek and Roman for nonconformist chapels, Moorish for warehouses, Indian Moghul for law courts, Italian Renaissance for breweries, all these styles met and mingled in the metropolises of the mid-nineteenth century. Nothing was barred: malted milk was made in a factory with a medieval castellated exterior, jails were built to resemble medieval keeps, and other popular styles include Byzantine monastic, Italian palazzo, French chateau, German baronial, Flemish Renaissance, and even Japanese. In the centres of large Canadian cities and in lonely spots deep in the Rocky Mountains are to be found huge nineteenth-century hotels built in an attractive combination of French chateau and Scottish baronial styles. It was an exuberant era.

Such wild Victorian exuberance, of course, meant a great reduction in the previously prevailing architectural uniformity and harmonious elegance of townscape. Some of this was recaptured, after the 1890s, via the Garden City movement and the subsequent development of vast suburbs in the twentieth century (Howard 1898).

By the mid-twentieth century, however, city cores were increasingly dominated by tall steel, concrete, and glass towers in the Modernist or Internationalist style (Wolfe 1981). Based on the technology of elevators and steel-frame building developed in the late nineteenth century, and on the socialist 'form follows function' ideology developed by the Bauhaus group in interwar Germany, the new style eschewed ornamentation almost entirely. Ironically, the style was taken up most avidly in the capitalist United States, whence it has been exported the world over. Soon, early 'Gothic' skyscrapers, erected in the early twentieth century from Chicago to Moscow, were to be dwarfed by enormous cuboid buildings erected after the Second World War in the Modernist style (Figure 2.28).

The general aesthetic effect of these buildings varies with viewpoint. At a distance of several miles, the effect of a wall or pyramid of tall towers in a city core can be stunning. It harks back, of course, to the monumental centralism of ancient cities, and reminds us that we are confronting yet another Ozymandian landscape of power. This is, however, now corporate power, for rarely do governments build structures to overtop or even match the tall glass towers of multinational corporations and finance houses.

Closer in, the effect is to dwarf the observer, as well as to bore her (Figure 2.29). Cuboid forms, flat roofs, flat façades, identical fenestration for forty floors, backs which look like fronts, white, grey, or concrete-coloured walls, and a minimum of decoration produce an effect of clean, clinical, soulless efficiency. This may, indeed, be the effect desired by banks and corporations. To the onlooker, after the initial glance, it is aesthetic death.

The proliferation of concrete is one major problem. Unlike brick or stone, concrete does not mellow. Instead, it rapidly becomes dirty, dark, and stained, reflects even less light than before, and soon takes on a dull and dingy look. From a wider point of view, the proliferation of concrete and glass towers has irremediably altered the cherished skylines of such cities as London and Paris. Until 1952 central London enjoyed a height ceiling of one

Figure 2.28 Modernist streetscape, Vancouver, BC

hundred feet; its skyline was a traditional one of spires and domes. The massive growth of office towers in the last half century has effectively eliminated this characteristic skyline, so that from several viewpoints London is now beginning to look like Dallas.

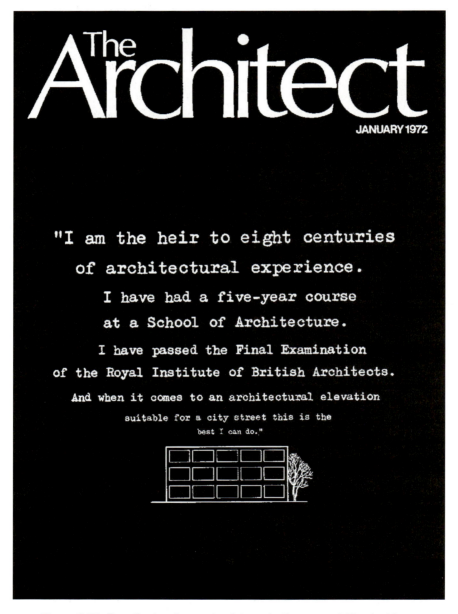

Figure 2.29 Questioning the modernist aesthetic: cover of *The Architect*, January 1972 (courtesy *The Architect*)

Figure 2.30 Standardized post-modern architecture: Songhees area, Victoria, BC, already known as 'Olde Nowhere'

The inevitable reaction to this brutalism was the Post-modern movement (Jencks 1978). Based on notions of pop architecture popularized by Venturi (1966), and on concepts of architectural allusion, exoticism, and playfulness, this movement developed in the late 1960s, and in the 1970s a number of interesting, colourful buildings appeared. By the 1980s, however, architectural Post-modernism had become the norm for new building, endlessly reproduced in North American cities in the form of tacky pediments, columns, arched windows and the like (Figure 2.30). This rapid degeneration of a new style into what has been called 'cornflake-packet architecture' and the 'Disneyland style' is a sad comment on the late twentieth century's inability to produce a pleasing and elegant architectural style which can be varied in each locale according to local requirements, resulting in a distinctive townscape. The 1990s has already begun to witness nostalgia for the clean-cut structures of Modernism. These issues are explored further, in terms of urban aesthetic planning, in Chapter 5.2.

2.4 SOME NATIONAL CONTRASTS

In the previous pages I have investigated fluctuations in landscape tastes through time. The landscapes themselves, as objects, have also been analysed. It is time now to look at variations in landscape appreciation across space. English and American tastes are first compared and contrasted, with brief subsequent accounts of landscape taste in Australia and British Columbia, Canada.

English landscape tastes

Few critics have so neatly encapsulated the essence of English landscape tastes than the geographers Lowenthal and Prince (1964, 1965). The basic theme in the English taste for landscape is the bucolic; the English are wont to speak of 'countryside' rather than 'scenery' or 'landscape'. The poet Cowper's notion that 'God made the country and man made the town' is historically inaccurate but spiritually correct. Village life is sanctified and there are attempts to countrify the city via parks, turf, and trees.

The bucolic includes both the pastoral, which involves open, rolling grassy downs complete with sheep, and the rustic, which is essentially arable land with trees. The trees must be rounded, bushy-topped, broad-leafed, and deciduous; there is a distaste for 'German conifer-worship'. Coniferous trees, a Western North American staple, are rejected by the English on the grounds that they lack variety in colour and shape, permit little undergrowth, show little seasonal or other temporal change, and have hard outlines. It took the formerly-British author of this book about fifteen years to find beauty in the 'pointy trees' of his British Columbian homeland.

The rolling smooth greenness of the pastoral downland symbolizes the English tendency to prefer the tidy, neat, and kempt, best exemplified in the rash of best-kept village competitions and the once-ubiquitous signs 'Keep Britain Tidy' (not healthy, safe, or lovely!). This emphasis on outer appearance is termed 'façadism' by Lowenthal, and it is especially important in historic townscapes where appearance is more highly regarded than convenience.

In rural areas, these attributes coalesce in a strong preference for the beautiful and the picturesque. English tourists, who visit 'stately homes' and parks in droves every summer, are voting with their feet in favour of a tamed rural landscape, a landscape wherein nature has been modified to appear as art (Figure 2.31). This emphasis on the elegant pastoral is varied by a strong liking for the more irregular, complex, and intricate landscapes of the picturesque, composed of thatched cottages, rutted lanes, and rookeries in elms (now sadly depleted by Dutch elm disease), the whole framed by a foreground of deciduous branches (Figure 2.32). Formal layouts and regimentation are eschewed; apparently overgrown and randomly-planted English country gardens are preferred over geometrically accurate tulip beds. The efforts of art are cunningly concealed behind an appearance of mild disorder.

Figure 2.31 English landscape tastes: the elegantly tamed
(Cowick Hall, Yorkshire)

Figure 2.32 English landscape tastes: the historic (church and moat
at Aughton, Yorkshire)

In keeping with this picturesque preference, medieval cities such as York are besieged by tourists seeking crooked lanes, irregularly shaped market-places, alleys, and courtyards; a small-scale, intimate, complex townscape no longer possible under modern town-planning standards. This mania for historic attachments may prevail to the point of retaining buildings for their historic value with no consideration of either their sensuous or functional values; 'history is worth having, even at the cost of comfort and beauty'.

Old things have associations with past events and people. The English are acutely aware that 'the past is another country' (Lowenthal 1985). Yet they are often unsatisfied with the bogus assemblages of historic artifacts that so please Americans. What such assemblages lack is a sense of place, the *genius loci* beloved of Pope and celebrated every few miles in traditional rural England in yet another difference in building or field-boundary style. The character of a place matters, and the aesthetic integrity of place is enshrined in national park regulations that confine the use of building materials to those available, and long used, locally.

The work of Lowenthal and Prince is the most prominent in a long line of aesthetic geographies that reach back to Cornish (1931, 1935) and Younghusband (1920). As one might expect in a survey of England, Lowenthal and Prince are sensitive to variations in preference associated with class, and they are strong in terms of conservation and preservation (Lowenthal and Binney 1981, Lowenthal 1985). Yet they fail to distinguish between the picturesque and the sublime, and make little distinction between eighteenth and nineteenth-century romanticism (Gibson 1989). Gibson's criticism that they pay little attention to the new British 'pop designs and tastes that were being formulated and implemented around them' is a little unfair in view of the publication of their seminal essays in the early rather than the late 1960s.

American landscape tastes

The American scene (Lowenthal 1968) stands in sharp contrast to the English. Like many American critics before him, Lowenthal is moved to judgement of his nation's landscape tastes. 'Face to face with the look of his own country, the well-travelled American is characteristically dismayed' (p. 61). Why is this so?

In essence, the American landscape is one of immense size, too large-scale by far for much of the beautiful or the picturesque to develop, especially beyond New England. All is sublime; landscapes are limitless, endless, vast, huge. People are dwarfed (Figure 2.33). Eighteenth-century visitors from Europe were overwhelmed rather than impressed. Americans, perhaps in reaction to European criticism, rejoined with a cult of bigness. Redwoods, the Grand Canyon, the claim that the Mississippi–Missouri was the world's longest river, huge automobiles, sixteen-ounce steaks, all these reflect a prevailing notion that bigger is better. This cult was translated into ever-taller skyscrapers in cities and into city planning via the dictum of Daniel Burnham

Figure 2.33 American landscape tastes: the vast (Arizona)

to 'make no little plans'. The best-known American structures are all monumental in scope, and they are admired less for their beauty than for their sheer strength and size.

The cult of wilderness is closely allied to that of bigness. England has no real wilderness, but the US retains wide areas of alien, impenetrable, unfathomable forest and mountain landscapes. Wilderness has entered the national consciousness, which tends to dismiss the tamed nature so typical of European landscapes as effete and feminized. The wilderness cult provides a home not only for the neo-Romantics of the Sierra Club but also for the legions of macho hunters, shooters, and fishermen for whom to penetrate the forest and maim its inhabitants, or one another, is a test of manly fortitude and strength.

With wildness and bigness comes a tendency to relish extremes. North America is well known for its extremes of climate and the sheer violence of natural hazards such as tornadoes and earthquakes. Pride is taken in the violent excesses of nature, as in the increasingly chaotic and untamable jungles of American cities. These metaphors are deliberate and in common use.

Whereas England at least tries to be tidy (increasingly a losing battle), America is fundamentally formless. Landscape transitions, so obvious and even legislated in England, are far less apparent here. Thick vegetation cloaks differences in terrain; we are often more aware of the uniformity of forest than of the difference between mountains, foothills, and plain. There is a complete lack of the European sharp edge between city and country. The cities themselves are in constant flux, and there is an endless process of rapid change. All this produces 'America's most distinctive look – casual chaos' (Lowenthal 1968: 69). It is not unexpected, then, to find books on landscape planning subtitled 'chaos or control' (Tunnard and Pushkarev 1963). And at the base of this ferment is an orientation alien to pre-Thatcherite Britain, the notion that the landscape should be worthy of its hire. Instrumentality is at the core of a prevailing view of the American scene, a view that is concerned with actively changing things and which often becomes a 'purely monetary view of landscape' (Lowenthal 1968: 81).

Other prevailing attitudes concern the American disdain for the present in favour of an idealized past and a glorious future in Tomorrowland. Those who live in the future are less likely to complain about an ephemeral (but strangely enduring) present landscape which appears to consist largely of flimsy, ugly, throw-away stage sets. The sudden growth of suburbs after 1945 and the more recent rapid outflow of urbanites to small towns and villages suggest that the small-town, picket-fence tradition has a strong hold on the American psyche. Yet the preferred past, unlike that of the English, appears to be a fake, 'history expurgated and sanitized' (ibid. p. 78), in the form of the replicas, reconstructions, and assemblages of real artifacts wrenched out of context, which are to be found in abundance in the Historylands and Old Sturbridge Villages of the nation.

Like wilderness, these historic constructs are fenced off from the everyday context, which is generally agreed to be not worth looking at. Hence the nearby and typical is neglected for the remote and spectacular, for only the unique, the big, and the special are regarded as worthy of interest. 'The features most admired are set apart and deluged with attention; the rest of the country is consigned to the rubbish heap' (ibid. p. 85). This approach Lowenthal calls 'featurism', and it is especially noteworthy in the many attempts to save individual historic buildings while neglecting the townscape in which they are embedded. Lowenthal puts this down to lack of local roots, a fluid social structure, little sense of heritage, a history of flimsy, disposable buildings, and a general feeling that 'it's surprising what you can get used to'.

These attitudes to space relate strongly to the excessive American compartmentalization of time. Just as American cities are segregated *de facto* according to race, class, and lifestyle, so the activity of looking-at-scenery is segregated to certain points in time. There are appropriate times for looking and seeing, and during these times 'scenic viewpoints' are visited, where the landscape is encapsulated into an interpretative statement and snapshot exposure data are often provided. In far less time than the eighteenth-century picturesque traveller took to adjust his Claude glass, the American tourist frames the view in his camera and clicks. 'The rest of the time', says Lowenthal, 'we are blind.' How far this judgement is accurate a generation later will be explored below in the Activists section (Chapter 4).

Australian landscape tastes

The first white Australian settlers arrived from Britain in 1788. With them they brought a taste for the bucolic and the beautiful. What they found was a kind of scratchy wilderness, lacking soft, gentle deciduous trees but abundant in harsh prickly plants, peeling eucalyptus bark, and an evergreen landscape of dull blue-greens and browns. The first reactions were, inevitably, that this wilderness landscape was both monotonous and frightening (Correy 1977).

A huge nostalgia for Britain emerged. Painters, unable to cope with the glaring light, painted the landscape as they thought it ought to look. Of course, it ought to have looked more English! Nature imitates art; nineteenth-century settlers avidly hewed down native trees, eliminating the 'weird melancholy' of the evergreen bush. The local flora was replaced by deciduous trees and by the flowers of the English country garden: roses, lupins, wallflowers, hollyhocks. Similar scenes took place, of course, in the hill-stations of the Indian subcontinent.

The general feeling at the turn of the century was that the Australian landscape was a barren and inhospitable wilderness: 'in the whole world there is not a worse country ... so very barren and forbidding that it may with truth be said, here Nature is reversed' (quoted in Correy 1977: 14). Although there were a few positive comparisons to English parkland or Romantic scenes, it was generally held that such a landscape would better be made over to look more like 'home'.

Before the mid-nineteenth century a deliberate attempt to tame the Australian landscape had been made. Fine houses were built, and landscape gardens created according to current English styles. Even Repton-style gardens were attempted. In the late nineteenth century Australian cities such as Adelaide were laid out as garden cities with broad avenues, formal squares, and broad parkland green belts.

Aesthetic appreciation of the native Australian landscape came slowly. By the mid-nineteenth century the rock formations of Sydney harbour were thought to have aesthetic value. Then the native vegetation, already conquered around the cities and therefore less feared, began to be regarded as something more than 'an enemy to be annihilated', and botanic gardens were set up. The positive views of visitors such as Mark Twain and Anthony Trollope had some effect, and by the 1880s Australian painters had begun to come to terms with their native landscape and to paint it not only as they saw it, but with love. By the 1930s the urban professional classes were developing an appreciation of wilderness through bushwalking, although this was not translated into an official national parks service until 1967.

Turning from the historic development of landscape taste to a probing of national values, Correy speaks with some heat of the so-called Australian Spirit. A petty-bourgeois mentality prevails in Australia; admiration is accorded chiefly to 'practical men' who 'make it' materially. Academics, poets, and dreamers are vulgarly disdained as 'bludgers' and 'wankers'. Thus both the designer of Canberra, the nation's capital, and the architect of the Sydney Opera House were forced to resign in disgust with the lack of aesthetic feeling of their clients, who appeared to be completely concerned with instrumental values and especially with money. Recent decades have seen a strong feeling in favour of megaprojects, a cult of bigness akin to that of the United States. Thus the Tasmanian government dammed the Gordon River and Lake Pedder, only to be embarrassed by a consequent excess of electrical power.

By the 1960s environmental protest had come to Australia. This, coupled with a florescence of Australian literature (Patrick White), painting (Sidney Nolan), and film, all of which celebrated the beauty, terror, and sublimity of the Australian outback, has led to some re-evaluation of attitudes, and even unions have been known to impose a 'green ban' on environmentally-sensitive construction work.

Landscape tastes in British Columbia

The development of landscape taste in coastal British Columbia closely parallels that of Australia. As with the arrival of the first colonists in what was to become the USA, we are dealing with an émigré mentality which carries with it a heavy load of cultural baggage.

The early explorers of the BC coast were not makers of culture but receivers of it, and as such their notions were rather the opposite of avant-garde. Thus despite prevailing concepts of the sublime, which the BC coast amply merits, the coast was first looked at largely in terms of the beautiful and the pastoral. As such, it proved 'dreadful' to the early explorers. Sir James Douglas, a mid-nineteenth-century governor, found only the lightly wooded area around Victoria to his taste. As for the rest, it was considered harsh, forbidding, and gloomy. And even later in the century, when Romantic notions had long been absorbed, mountains were seen as sublime largely in visual and moral terms; they were for looking at, rather than for living in.

These early views were relayed back to Europe, which consequently thought about the BC coast, when it thought about it at all, in rather negative terms. Settlers were as keen as the Australians to cut back the forest and replace it with English trees and flowers. Whole streets in the older parts of Victoria are lined with horse chestnuts, hawthorns, and other evidences of pre-1914 British botanical taste. The mission, as in Australia, was to banish the native and recreate 'home'.

This effort to establish England in Canada is aptly described in the autobiographies of Emily Carr (1966, 1978). Carr, BC's best-known twentieth-century landscape painter and a contemporary of the Group of Seven, was one of those responsible for converting white inhabitants of the coast from a negative to a positive attitude toward native landscapes. Tippett and Cole (1977) aptly consider this process as the elimination of an attitude of landscape as 'desolation', and its replacement by an appreciation of landscape 'splendour' (Figure 2.34). The novelist Malcolm Lowry was also

Figure 2.34 British Columbian landscape tastes: from 'desolation', through 'splendour', to 'beautiful' (BC Government photo)

instrumental, in the 1940s and early 1950s, in pointing out the sublimity of the coastal landscape to those who saw it chiefly in terms of mining or forest industry resources (Porteous 1992).

Since the 1960s the traditional 'BC Spirit', which like that of the United States and Australia tends to bigness and a worship of megaprojects, has been challenged by neo-Romantic wilderness and ecological organizations which call for increasingly large areas to be protected from civilized change. Despite the efforts of unions, large corporations, and petty bourgeois governments which are hand-in-glove with business interests, environmentalists seem to be becoming effective in preserving at least some of the more spectacular landscapes from the depredations of logging and mining. On a broader scale, a plan put forward by the World Wildlife Fund (Hummel 1989) seeks the preservation of a typical example of every landscape type in the whole of Canada, a mosaic of reserves which in aggregate would account for 12 per cent of the total land surface of the country.

Such plans, of course, must become public policy (Chapter 4) before they have any chance of being implemented. And in a scientistic culture like ours, their chance of becoming implementable public policy often depends on scientific support of the kind outlined next in Chapter 3.

3

EXPERIMENTALISTS

Easter Island *moai*

O the mind, mind has mountains; cliffs of fall
Frightful, sheer, no-man-fathomed ...

Gerard Manley Hopkins

It is difficult to describe the methods used by humanists, but close observation and careful interpretation loom large, coupled with deep reading in a wide variety of contexts. Of late, these historico-cultural methods have been dignified with seemingly more precise labels such as phenomenology and hermeneutics. Basically, however, humanists immerse themselves in their subject in a traditional scholarly way, and learn to think deeply and critically without the need for an extensive apparatus of technique. They accumulate data in an inductive manner, and are introspective and contemplative. Some become 'gurus' in their disciplines, with a reputation for wisdom now rare in university life.

Experimentalists are equally dependent upon observation and interpretation, but between the two they interpose some form of laboratory experimentation, field survey, or questionnaire approach. They are mostly psychologists and usually heavily wedded to a 'hard science' model of the scientific method, with all that this entails in terms of precision and accuracy in hypothesis-testing, research design, the calibration of research instruments, data generation and analysis, and a proper caution in interpreting results.

The largely quantitative experimentalist approach is explained (Chapter 3.1) in terms of the development of psychology's interest in environment and the kinds of questions that experimentalists ask. Experimentalists are interested in both environmental variables (such as complexity and mystery) and in the characteristics (culture, social variables, personality) of persons who interact with environment (3.2). Having an applied bent, some experimentalists would like to be able to demonstrate the truth of the ancient belief that 'scenery is good for you' (3.3). However, ever conscious of methodological problems (3.4), experimentalists frequently reserve their opinions with the useful academic phrase 'more research is needed'.

Taking the broad, and therefore the only permissible view, humanistic and experimentalist approaches are eminently complementary. Neither is essentially oriented towards the implementation of its discoveries, although humanists' ideas prompt activists (Chapter 4) and experimentalist techniques have been heavily imported into aesthetic planning systems (Chapter 5) as the basis for practical landscape evaluation.

3.1 THE EXPERIMENTALIST APPROACH

Experimental psychologists have studied aesthetics since the 1870s (Valentine 1962), but until recently their work was almost wholly concerned with art forms rather than landscapes. In a review of work accomplished by 1960 Dickie (1962, 1971) asserted that psychology was of little relevance to traditional aesthetics because: it had thrown little light on the meaning of aesthetic experience; it had thrown less light on the nature of aesthetic creation; and had shed still less light on the problem of assessing aesthetic worth. Dickie concluded that, in the assessment of aesthetic artifacts, we should fall back upon the opinions of highly sensitive, trained aesthetic experts renowned for their taste. In terms of the assessment of landscapes, a similar view has been expressed by some planners (see Chapter 5).

In a modification of Dickie's abandonment of psychology, Westland (1967) suggested that social science might have value to aesthetics in three possible ways. First, psychology's experience with perception studies might be valuable in further illuminating the mechanisms of aesthetic apprehension. Second, psychology might well be useful in investigating how people respond to aesthetic material. Social science may readily uncover consistent patterns among the preferences of, and judgements made by, persons viewing aesthetic *objets*. But although it is eminently scientific to quantify agreements of preference and judgement, such empirical work cannot answer the more fundamental questions that an aesthetician might ask, such as: What is the significance of these agreements and disagreements?

Nevertheless, scientific work since the nineteenth century has demonstrated that there is considerable agreement among tested subjects with regard to both their separate aesthetic judgements and statements of their underlying values (Prall 1929, Valentine 1962). On this basis, Westland felt that the third contribution of psychology would be the devising of tests to measure aesthetic feeling. The chief problem with such tests, according to Westland, was the lack of a standard against which individual differences in appreciation might be measured; an individual IQ, for example, is measured against statistical population norms.

One, albeit imperfect, response to this last problem has been to compare the responses of experimental subjects against the average response of art experts. Results of a series of investigations reported by Child (1962) indicated that experimental subjects unselected for their knowledge of art tended toward high levels of agreement in their art preference ratings; this tendency toward agreement was consistent across all kinds of paintings (e.g. religious, group, still life, landscape) except abstract art. Nevertheless, experts asked to assess the 'aesthetic value' of the paintings, rather than merely express preferences, agreed with each other decidedly more than did the experimental subjects.

Indeed, there were only very low correlations between subjects' preferences and experts' evaluations, suggesting that what the public expresses as its preferences has little to do with what art experts regard as aesthetic value.

In view of the emphasis placed on public preferences by some experimentalists and planners, this is a most important finding. We shall find in Chapter 5, below, a fundamental disagreement among landscape evaluation professionals concerning the issue of who shall evaluate landscapes, experts or the public. Only where 'the public' can be defined as a highly-educated, cultured subgroup can any major degree of agreement between the two be expected. As Child (1962) reports, public preference agreements with expert evaluations correlated positively with the subjects' knowledge of art and scores on the scholastic aptitude test.

Much of this research was geared toward the development of measures of aesthetic sensitivity, defined as the extent to which a subject's preference judgements about visual art agree with those of art experts. It was consistently found that aesthetic sensitivity is a very stable characteristic of the individual, relating strongly to personality variables such as cognitive independence and openness (Child and Iwao 1968), and without any necessary relationship to a passive acceptance of convention.

In terms of environmental aesthetics, these emphases on individual differences and the public/expert dichotomy should be set against the large-scale generalizations of both humanists and planners. It is possible, however, that judgements of works of art may be qualitatively different from judgements of real landscapes, an issue which has received little attention.

Summarizing the issue of art preference studies, Westland (1967) concluded that psychology was unable to deal with the fundamental value questions of aesthetics which, being immeasurable, must perforce be left to the attention of humanists. 'Questions of fact', however, such as the study of public preferences, are measurable, and as such 'should be measured as vigorously, as objectively, and as unambiguously as possible' (p. 356).

Psychologists and environment

As we have seen, early psychological aesthetics focused largely on works of art. Indeed, when Maslow and Mintz (1956) began their classic studies in the 1950s, they found a dearth of background research on which to build. Almost all of the earlier work on colour, music, art and the like had used fragmented environmental components (e.g. patches of colour, works of art) as stimuli; the complex molar environment had been largely ignored. Further, they found no previous work in experimental psychology on the effects of environmental beauty or ugliness upon people.

Maslow and Mintz's classic experiment involved the creation of 'beautiful', 'average', and 'ugly' rooms. Subjects in these rooms were asked to judge photographs of male and female faces on 'energy' and 'well-being' dimensions. Typically, the same face would be judged, in the beautiful room, to be high in energy and well-being, whereas when judged in the ugly room it would be seen to display more fatigue and displeasure. When the test administrators were themselves tested (Mintz 1956) it was found that over the long term these findings were even more apparent, for the examiners in the ugly room complained more often of monotony, headache, sleeplessness, irritability, hostility, and discontent. The modest conclusion to this set of experiments was that 'visual-aesthetic surroundings ... can have significant effects upon persons exposed to them'. Psychology, as we will discover, frequently confirms what humanists and designers have intuitively known for millennia.

Maslow and Mintz's study has achieved classic status in psychology and building design (Brebner 1982, Russell and Snodgrass 1982), and several confirmatory replications have been performed. Both Locasso (1988) and Wilmot (1990), however, failed to replicate the findings of Maslow and Mintz in several studies, and report other work of similar tenor. Wilmot used actual room settings, whereas the earlier replicators used photographs, slides or verbal descriptions of rooms, a methodological issue considered more extensively below in Chapter 3.4. Perhaps because the aesthetic impact of environments is now seen as a much more complex phenomenon, mediated by cultural factors and personal experience, recent work on the effects of interiors has downplayed aesthetics and focused on work-related issues such as the relationship between lighting and task performance (Nasar 1988, Veitch 1990, Veitch, Gifford and Hine 1991).

The next step, and a difficult one for a discipline wedded to a rigorous, hard-science approach, was to come out of the laboratory and into the street. In fact, much of the early work on the perception of molar environments was performed by workers in applied fields such as city planning and forest management.

In the former, the work of Lynch (1960) was seminal. In his classic study 'A walk round the block', Lynch (1959) asked the basic question: 'What does the ordinary individual perceive in his landscape; what makes the strongest impression on him and how does he react to it?' In a naturalistic experiment walkers were asked to tell the experimenter about the things they saw, heard, or smelled. Although the main aim was to discover how people created mental order from their impressions (and which led to decades of research on mental maps), a secondary focus of interest soon emerged because of the spontaneous aesthetic remarks made by the subjects.

Spontaneous emotional responses such as 'it's lovely' or 'it's ugly' were commonplace, and pedestrians commonly expressed enjoyment of the openness of wide sidewalks, a dislike of the strict enclosure of narrow alleys, and enjoyed the loveliness of buildings lapped in sunlight and the rich texture of sandstone. Some found church steeples to be exciting punctuation marks; many found the profusion of signs confusing rather than informative.

More strictly controlled experimental work was simultaneously taking place in wilderness settings under the aegis of forest managers and related professionals. Like the city work of Lynch, these investigations were concerned with orientation, perception, and activity (Lucas 1964). As with the study of urban perceptions, however, it was soon discovered that aesthetic satisfactions were an important aspect of wilderness users' perceptions. The classic study of Shafer and Mietz (1969), for example, demonstrated that aesthetic satisfactions (e.g. the contemplation of scenic beauty) and emotional satisfactions (e.g. reaching the top of a peak) were far more important to wilderness hikers than physical (exercise), educational (learning) and social satisfactions. These findings have been confirmed again and again in a plethora of recreational research both quantitative (Lucas 1970a, b, Marsh and Gardner 1978) and qualitative (Porteous 1991).

It is clear that experimentalists must consider the nature both of environment and of those who observe it. It is logical to take an interactionist viewpoint and state that the quality of beauty is not exclusively a property of the environment, nor yet exclusively in the eye of the beholder, but depends on the necessary interaction of observer with environment.

Psychology, however, is nothing if not analytical. In practice, therefore, much of the research in experimental environmental aesthetics has concentrated on one or other of two basic questions: what are the environmental determinants or components of aesthetic response; and what are the personal determinants or components of aesthetic response. In other words: (1) what is it about environment that disposes us to react to it aesthetically?; and (2) what is it about ourselves that influences our aesthetic reactions to environments? As the following subsection demonstrates, research work varies in the emphasis which is placed on environment or person.

3.2 PROPERTIES OF ENVIRONMENTS AND PERSONS

Psychological work in environmental aesthetics has tended to consider either environmental variables or attributes of the person, but, until recently, rarely both.

Environmental variables

Psychologists concerned with the environmental characteristics that appeal to our aesthetic sensibilities have tended to approach the subject in two distinct ways, which may be termed (1) the Berlyne–Wohlwill approach, which seeks to establish the nature of 'collative variables', and built upon it, (2) the Kaplan approach, the main thrust of which is to develop a model of environmental preference.

The Berlyne–Wohlwill approach

This approach is grounded in a theoretical background based on Berlyne's *Conflict, Arousal and Curiosity* (1960). Berlyne (1971, 1974) was interested less in aesthetics than in exploratory activity. His general hypothesis, verified by a large body of experimentation, was that individuals will engage in the voluntary, active exploration of a stimulus in direct proportion to the amount of uncertainty and conflict it engenders.

Two types of exploration are envisaged (Wohlwill 1976). 'Specific exploration' occurs when an individual is confronted with a stimulus which generates conflict or uncertainty; the individual usually seeks to lower a high arousal level by exploration. Conversely, 'diversive exploration' occurs when an individual with a low arousal level (e.g. when bored or tired) seeks some stimulus which is not too familiar, yet not too novel (e.g. a seldom-watched television programme) by means of which some higher but optimum level of arousal may be attained. It appears that hedonic value (aesthetic satisfaction) is related to stimulus uncertainty according to an inverted-U-shaped function very similar to that uncovered by Berlyne's studies of exploratory choice in the diversive exploration mode.

The resulting Berlyne–Wohlwill schema of aesthetic response to environment is expressed in Figure 3.1. Briefly, the environment may be viewed as a vast array of collative variables, so called because these variables are essentially comparable across environments; in order to decide how novel or complex a stimulus pattern is, for example, one must compare or collate information from two or more sources. Based on the collative variables, stimulus patterns in the environment have the attribute of arousal potential. Actual arousal in an observer will depend on how alert that person is at the moment of observation. This arousal may result in the achievement of hedonic value, the pleasure or reward obtained from the viewing of a work of art or landscape.

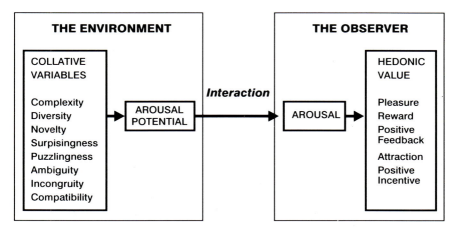

Figure 3.1 A schema of aesthetic response to environment
(after Berlyne 1971, 1974)

In operationalizing this theory a wide variety of verbal, behavioural, and physiological measures have been used to record aesthetic responses. Most typically used are seven-point semantic differential scales, where a relaxed-tense spectrum might be used to measure arousal and ugly–beautiful/ pleasant–unpleasant scales for measuring hedonic value. The characteristic inverted-U-shaped pattern of response (Figure 3.2) has been demonstrated in many experiments (Chalmers 1978). Typically, as arousal potential increases, hedonic value also increases until an optimum point is reached. Thereafter, further increases in the environment's arousal potential are followed by a downturn in hedonic value. Wohlwill and Kohn (1976) call this the 'optimization principle'.

Hedonic Value

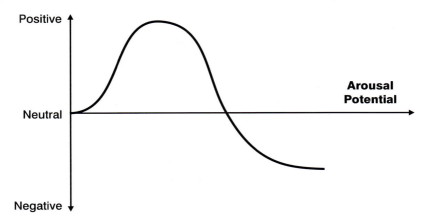

Figure 3.2 Characteristic pattern of aesthetic response to increasing levels of arousal potential (after Berlyne 1971, 1974)

Clearly, this approach to the aesthetic appreciation of environment must pay particular attention to the collative variables which generate arousal potential. The collative variables which have received most attention would not be wholly surprising to some classical theorists nor yet to eighteenth-century landscape designers; they include complexity, diversity, novelty, coherence, surprisingness, puzzlingness, ambiguity, incongruity, incompatibility, and a host of others of lesser importance.

Aestheticians and designers have frequently emphasized the importance of complexity, especially when linked with ambiguity and coherence. The importance of complex ambiguity has been emphasized in literature, architecture (Venturi 1966, Herzog *et al.* 1976) and the social sciences (Rapoport and Kantor 1967). Unfortunately, experimentalist results are themselves ambiguous on the scores of both complexity and ambiguity.

An array of studies of rooms, buildings, shopping centres, cities, natural landscapes and others shows great variation in the correlation between respondents' ratings of complexity and their ratings of beauty, and Wohlwill (1976: 50) concludes that 'complexity ... plays an uncertain role at best in the individual's aesthetic response to the environment'. The results from studies of ambiguity are equally problematic, not least because of the lack of standardization of the concept, which closely connects with what researchers have also called 'uncertainty', 'incongruity', and 'compatibility'.

The notion of compatibility, or fittingness, however, has proved an important variable, with several studies revealing strong correlations between verbal measures of compatibility (i.e. how different land uses and landscape elements 'fit together' in the scene presented) and judgements of beauty (Steinitz and Way 1969, Wohlwill 1976, Chalmers 1978). Indeed, the 'index of compatibility' developed by Hendrix and Fabos (1975) appears to have practical use in predicting public judgement of landscape change, and Zube, Pitt and Anderson (1975) found it the most useful of a whole array of measures designed to assess scenic resource value (see Chapter 5.1).

The Kaplan approach

A very large body of experimental work in environmental aesthetics has been performed since 1970 by R. and S. Kaplan; much of it is summarized in their *The Experience of Nature* (1989). A major thrust of this research programme is the development of a model of environmental preference. For the Kaplans, preference is not simply a liking for one setting over another; preference actively guides behaviour and learning (S. Kaplan 1983). The theoretical base for their model is the assumption that preference relates to the individual's ability to acquire information so as to make inferences about his or her whereabouts. During the early evolution of the human species, certain types of environments proved more useful for survival than did others, and these have become the preferred environments of *Homo sapiens*. To this point the Kaplan approach is consistent with Appleton's (1975a) habitat theory and the work reported in Chapter 1.3 above on the human preference for savannah landscapes.

The Kaplans go on to suggest that those properties of environment which satisfied the informational needs of our primitive forebears now serve as determinants of modern environmental preferences. Six major variables are said to influence preference most strongly. They are: complexity, mystery, coherence, texture, identifiability, and spaciousness (R. Kaplan 1975). Not surprisingly, these variables coincide to a great extent with Berlyne's collative variables and with some of the technical terms used by eighteenth-century landscape gardeners. The methods used are not unlike the Berlyne–Wohlwill approach; slides or photographs are shown to experimental subjects who rate their preferences on semantic differential scales. Preference is operationalized as 'how much do you like this scene?' Scenes are then grouped into 'context domains' by scaling or cluster-analysis, and then analysed in terms of predictor variables by regression or correlation.

The Kaplans' work supports the humanists' and designers' belief in the importance of complexity, coherence, and ambiguity, which the Kaplans more usefully call 'mystery'. There are strong correlations between preference and all these variables, although texture, spaciousness and identifiability do not score so well. Perhaps the most interesting results concern mystery, defined operationally as 'the promise of further information based on a change in the vantage point of the observer. Consider whether you would learn more if you could walk deeper into the scene' (R. Kaplan 1975).

Both in art and in townscape and nonurban landscape design the role of mystery has been asserted. In relation to the Kaplans' confirmation of this, however, Wohlwill (1976) argued that there is surely some inconsistency between the Kaplans' hypothesis that preferred scenes will be high in both mystery and identifiability (legibility). Ulrich (1979a) responded to this problem by factoring in the time-element, arguing that legibility relates to present information, the ease with which the parts of a scene can be organized into a coherent pattern at this moment. Mystery, by contrast, relates to future information possibilities, the promise of more should we enter the scene and explore it. The Kaplans (1982) have built on this and their earlier work (S. Kaplan 1979) to provide a preference framework (Table 3.1) which views preference as a need both to make sense of a scene and to be involved in it. Coherence permits a quick understanding of the scene, while

Table 3.1 The Kaplans' preference framework

	Needs	
Availability of information	*to make sense*	*to be involved*
Immediate (present)	Coherence	Complexity
Promised (future)	Legibility	Mystery

Source: adapted from Kaplan and Kaplan (1982)

complexity offers enough information to promote interest. Legibility reduces feelings of disorientation, while mystery encourages exploration (Figures 3.3, 3.4). It is hypothesized that highly-preferred scenes will contain

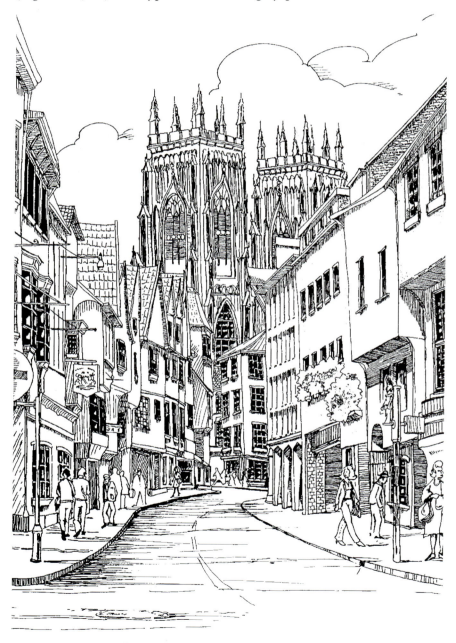

Figure 3.3 Coherence, complexity and mystery in York, England

Figure 3.4 After an uphill climb to the gate, mystery yields to the reward of a complex, coherent view of rural County Durham, England

all these elements, but that preference will also depend in an extremely complex manner on sociocultural variables (see below).

Ulrich (1979a) then attempted to reconcile the Berlyne–Wohlwill and Kaplan results, restating these as follows: visual landscape preference is a response in favour of scenes (a) which effectively transmit landscape information, i.e. are legible, and (b) which convey a sense that more information can be gained at low risk, i.e. mystery. Legibility, in Ulrich's terms, has four components: complexity; focality (coherence, unity); ground texture; and depth (which, of course, is the factor which permits the exploration promised by the element of mystery). Ulrich also suggested that two kinds of complexity should be considered, the patterned complexity (more legible although complex) of the Kaplans and the random complexity (lacking structure or coherence) of the Berlyne–Wohlwill model. This, he felt, would account for the different results obtained by the two groups when studying complexity.

Testing his model of preference on Swedes and Americans, Ulrich found that both had similar preferences, with high correlations between preference and focality, ground texture, depth, and especially mystery. Indeed, preferred scenes were characterized by even and fairly homogeneous ground texture, medium to high levels of depth, the presence of a focal point, and moderate levels of mystery, exactly as predicted. For neither group, however, was preference related to complexity, although this may have been due to the unfortunate failure to distinguish experimentally between patterned and random complexity. Incidentally, the most preferred scenes were smooth, grassy, parkland settings with background trees, which transports us straight back to the evolutionary savannah theory of aesthetic preference (Chapter 1.3).

Personal variables

The aesthetic judgements made by experimental subjects are not responses initiated by the environment alone. Subjects are clearly influenced in their responses by a complex array of personal characteristics and attributes. Dearden (1989) conceptualizes these influences as a nested hierarchy (Figure 3.5) wherein the most basic, or 'innate', level, has already been discussed in terms of theory development in Chapter 1.3.

Culture

Cultural differences in landscape taste have also been dealt with at length, for this area has been paid considerable attention by humanists (see Chapter 2

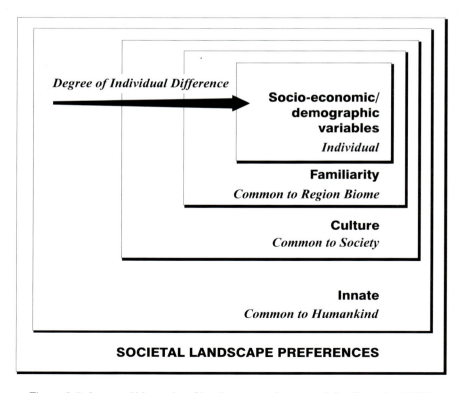

Figure 3.5 A nested hierarchy of landscape preferences (after Dearden 1989)

and, e.g., Tuan 1974, 1977, 1984, 1993). In comparison, the experimental evidence for cultural differences in aesthetic perception is relatively sparse (Cole and Scribner 1974). Nevertheless, pioneer experimental work has been useful in pointing out major differences in aesthetic preferences between dominant and subordinate groups, such as American tourists and Jamaicans (Zube and Pitt 1981), non-Inuit and Inuit (Sonnenfeld 1967), American whites and blacks (Flaschbart and Peterson 1973), and residents and visitors (Orland 1988). Specific attributes such as urban residential quality (Zube, Vining, Law and Bechtel 1985) and the 'aesthetic incongruity' of billboards and litter (Sorte 1971, 1975) have also been found to have a cross-cultural dimension. Other studies, however, have found strong similarities between the aesthetic preferences of Americans and Australians (Kaplan and Herbert 1988), Swedes (Ulrich 1979a) and Japanese (Nasar 1984).

The native–non-native dimension, with ample evidence that natives most often prefer scenes of their native habitats, has received considerable attention (Sonnenfeld 1966, Lyons 1983), although other research has questioned the universality of such findings (Canter and Thorne 1972, Ulrich 1979a, Nasar 1984). Indeed, a scenic beauty evaluation study found strong similarities in landscape preferences between native Balinese and Western visitors to Bali (Hull and Revell 1989). The authors speculate that this could be the result of genotypical preferences for certain landscape types, especially parkland with views (the savannah theory) or of innate preferences for landscapes with certain levels of complexity, focality, texture, mystery etc. (the Berlyne, Wohlwill, Kaplan and Ulrich approach). They show that scenic beauty values are indeed learned from the culture, at least among the Balinese, and then wonder how far native Balinese aesthetic preferences have been overlaid by the spread of Western values with the coming of mass tourism.

If the issue of culture is so contentious, that of familiarity is a little clearer. Using the concept of adaptation-level theory (Gifford 1987), Flaschbart and Peterson (1973) suggested that individual preferences are strongly influenced by past experiences, and Herzog, Kaplan and Kaplan (1976) found familiarity to be an effective predictor of preference. These notions relate to what Acking and Sorte (1973) call 'affection', and the concept of attachment discussed above in Chapter 1.1 and later in 5.4. Memory is obviously involved here. Herzog et al. (1976) found no difference between subjects' ratings of pictures of urban scenes and ratings based simply on the corresponding place-names. Merrill and Board (1980), in a series of experiments, showed that subjects judging familiar surroundings from memory gave very similar ratings to those given by subjects making judgements in the field. Preference ratings, then, do not seem to fall off even when no immediate visual input is available.

The native–non-native work as well as urban image studies (Lynch 1960, Appleyard 1969, Porteous 1977) have found that the familiarity of respondents with the actual sites being evaluated, or with similar sites, is a major factor influencing landscape preferences (Beckett 1974, Jackson, Hudman

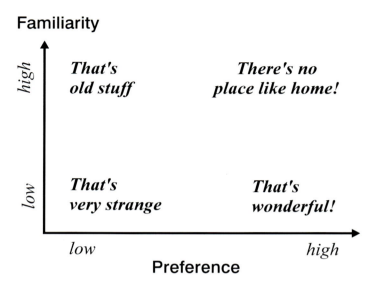

Figure 3.6 A familiarity and preference matrix (based on an idea in Kaplan and Kaplan 1982)

and England 1978, Nieman 1980, Clamp 1981). Familiarity-related preference may change significantly with scale and over time (Dearden 1989), and the Kaplans' (1982) 'familiarity and preference matrix' (Figure 3.6) illustrates their notion that life is a continuous trade-off between the 'excitement of the new and the comfort of the known'.

Social variables

Socio-economic and demographic variables have received considerable attention from experimentalists. Age appears to be extremely significant. A number of studies report that children under twelve are less discriminating than adults and show much greater variability in response to landscapes, have little interest in high naturalism in landscapes and are less likely to view human intervention in the landscape as detrimental (Zube, Pitt and Evans 1983). This accords with parents' observations that their children only begin to be socialized into the appreciation of scenery after about age nine (Allen 1989).

In detail, the work of Zube *et al.* (1983) suggests that there is high agreement between adolescents, young adults and the middle-aged on scenic values, though the responses of both the very young and the elderly are quite different from this norm. Whereas neither children nor the elderly seem to be strongly influenced by human influences in natural scenes, adolescents and adults tend to judge scenes as of ever lower value as the man-made component increases. Adults were far more sensitive than children to landscape form and land-use compatibility, whereas children seemed far more sensitive than adults to water, a finding which does not seem to match the work of Ulrich discussed below in Chapter 3.3.

It is possible to cite works that find age to be of little importance and others which find it to be the most important predictor of preference (Lyons 1983). As an experimental variable, age, like complexity, is too often considered too simply. Most studies assume that preferences will vary with age as the total environmental context changes or because of the aging process. Little work has been done on the hypothesis that there may be intergenerational differences in landscape values (Dearden 1989). More light might well be shed on the subject if more attention were given to studies of childhood aesthetics. For a discussion of this issue, see my work on 'childscape' (Porteous 1990).

Lyons (1983) found that age and gender variables combined to form a powerful predictor of preference. Insufficient work on gender, however, prevents firm conclusions being drawn, other than that considerable differences have been found in some studies (Macia 1975, Dearden 1984) and that females appear to show higher appreciation of both periurban and urban scenes (Peterson 1975). The issue deserves more attention if only insofar as most landscape designers and environmental managers are middle-class, middle-aged males (see Chapter 5.4).

Socio-economic class-related variables, such as occupation and expertise, have a considerable influence on landscape preferences. Early work focused on subjects' instrumental relationships with environment, and stressed the differences in environmental perception and cognition between 'special competence groups' such as planners, architects, and space-managers, and their clients (Craik 1968). As with the age and gender variables, some studies have found considerable differences in outlook between various types of experts (R. Kaplan 1975) while others have revealed no such differences (Dearden 1984).

R. Kaplan (1973b), for example, found significant similarities in the importance of environmental attributes as predictors of preference between architects, landscape architects, and other college students. For all groups, the coherence (legibility) of a scene was a strong predictor of preference, although architects perceived most coherence in photographs of building complexes, while other college students saw most coherence in natural scenes. In terms of stated preferences, architects gave highest ratings to photographs of buildings without nature and lowest ratings to scenes of nature only. Landscape architects, in contrast, strongly preferred mixed scenes of buildings with nature, while general college students' preferences reversed those of architects. Kaplan concluded that whereas the importance of environmental attributes such as coherence as predictors may vary little between groups, strong differences in scenic preferences exist as a function of professional interest. This could well lead to problems in team projects, although Zube (1974) has shown, in a comparison of environmental designers (architects, landscape architects and planners) and resource managers (foresters, hydrologists, wildlife managers), that 'there is a reasonable level of agreement on a set of evaluative and descriptive terms relating to scenic resources'.

But what of the gap between expert and public? (an issue explored in detail below in Chapter 5: Planning). During the later twentieth century, for example, architects have assumed that the general public accepts the distinction between modernist and post-modernist buildings and finds the post-modern style more meaningful. Testing this hypothesis, Groat (1982) confirmed earlier work on the expert/public dichotomy by discovering that architects and accountants (surrogates for the public) made judgements about buildings based on quite distinct, although overlapping, sets of criteria. Further, only the architects drew the modern/post-modern distinction; for the accountants it was not a relevant mode of distinguishing between buildings. This rather suggests that Tom Wolfe (1981) is right in his belief that post-modernism in architecture is merely a variation on the prevailing modernist style.

Personality

Further work on individual differences awaits the elaboration of techniques capable of measuring individual dispositions towards the environment. Rachel Kaplan's (1977a) Environmental Preference Questionnaire is such an instrument, and has led to generalizations based on the assumption that environmental preferences are an enduring, rather than a fleeting, property of individuals.

For example, Kaplan makes an interesting distinction between those who prefer to live in suburbs and those who prefer to seek out nature. The suburbanite likes to achieve, accomplish, and lead. He feels himself to be realistic, confident, and important, and enjoys competitive activities involving other people. Tranquillity is not important. In sharp contrast, the nature-seeker likes to explore and feels confident in her ability to do things on her own. Peace and quiet are important, although the enjoyment of solitude may be shared by joining in with small groups engaged in cooperative rather than competitive activities.

Personality is also important in relationship to dense urban surroundings. A complex research investigation by Janssens (1984) in Sweden first confirmed that the evaluation of a building as pleasant related to its visual complexity. Extroverts were then found to perceive buildings as more complex than did introverts; when complexity is needed, remarks Janssens, it can be found. Neurotics perceived tall, massive buildings as unpleasant, whereas persons with a strong sense of self perceived them to be smaller, less massive, and less complex.

On this basis further work might confirm the existence of 'environmental personalities'. In the cases noted above we see a contrast which may relate to the common distinction between Type A and Type B personalities. Such work may well be used to predict environmental activities, patterns of satisfaction, and even developmental trends in careers. We might also expect differences in commitment to the ecology movement, to 'green' activities, to war-making and Third World development aid.

Further research in this interesting arena has explored the possibility of bringing personality to the fore in environmental psychology research (Craik 1976), with a view to the elucidation of environmental dispositions. Little's (1976) study of specialists revealed the possibility of four major specialist groups (Figure 3.7) of which the non-specialist and person-specialist tend to have little interest in environment while the thing-specialist approximates to my Experimentalists and the generalist to both Humanists and Planners. Using McKechnie's (1974) Environmental Response Inventory (ERI), Phillips and Semple (1978) surveyed first-year students in the University of Waterloo's Faculty of Environmental Studies (ES), which comprises four units: architecture (ARCH), geography (GEOG), man–environmental studies (MES) and urban and regional planning (URP). The ERI measures environmental personality on nine scales: pastoralism, urbanism, environmental adaptation, stimulus-seeking, environmental trust, antiquarianism, need for privacy, mechanical orientation, and commonality. Seven personality clusters emerged (Table 3.2).

The significance of the existence of such environmental personalities, if confirmed by further studies, is profound, and has considerable import for the discussion of Activists (Chapter 4). For while few would want their world controlled by spectators and negatives, it is clear that many decisions

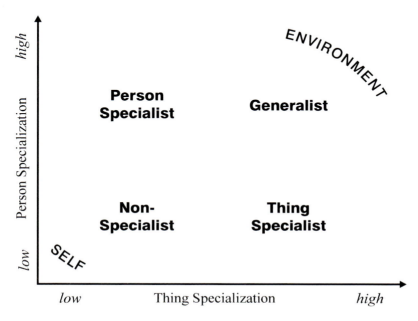

Figure 3.7 Four primary specialist groups (after Little 1976)

Table 3.2 Phillips' and Semple's environmental personality clusters

Group	Characteristics	Personality	Discipline
Unemotional technocrats	little concern for natural environment, prefer urban. Need familiar security. Low on pastoralism, communality, stimulus-seeking, antiquarianism	urban, intellectual, conservative, apathetic, restricted, blocked affect, critical, self-controlled	32% of ARCH 28% of URP Few GEOG or MES
General academics	very urbanist, very low need for privacy, high scores on adaptation, trust, and communality; environment is a resource to be used; faith in technology	urban, gregarious, socially adroit, intelligent, cultured, controlling, rule-following, planning	35% of URP few MES
Spectators	low urbanism, high stimulus-seeking; uninvolved in environmental transactions	spectator; unsure of goals and values	35% of MES few URP
Negatives	low on all scales; reject environmental influences; actively block environmental stimulation	overcontrolled, self-righteous, fearful, conscientious, isolated	16% of GEOG few ARCH
Environmentalists	high on all but urbanism, trust, and adaptation; appreciate peace and solitude of nature, artifacts from cultural past; stereotypical ES student	artistic, self-willed, high in affect	11–15% of each unit
Eco-freaks	reject urban, worship the natural with almost religious fervour; uncompromising stand against any modification of environment	vary from warm, appreciative, emotional, dreamy, to coarse, cold, intolerant, prejudiced	18% of MES 12% of GEOG
Conservationists	like eco-freaks but without emotionalism and extremism. Emphatic, ecological concern for natural environment; confident, self-sufficient nonconformity	hard-headed, managerial, versatile, opportunistic, polished, psychologically-complex	16% of MES 13% of ARCH

Source: data derived from Phillips and Semple (1978)

to change environments are made by unemotional, technologically-oriented planners and architects, who lack the commitment of the more generalist eco-freaks and conservationists. Thus those who have a rich and flexible generalist outlook, one most likely to contribute to a balanced approach to environmental change, find themselves powerless in the face of the rule-governed, group-cohesive, elitist professionalism of the planners.

Environmental disposition work continues with the study of wilderness purists (Graber 1976), work on 'the tour personality' (Wood and Beck 1990), and studies of the development of personal environmental schemata (Amedeo and York 1990).

It is here that the artificial division between work on 'objective' environmental attributes and 'subjective' personal characteristics comes together in the person–environment–behaviour framework which is now becoming the accepted model for research in environmental aesthetics (Dearden 1989, Kaplan and Kaplan 1989). It is likely that future research orientations will change dramatically from the past 'either-or' approach, 'given the very widespread awareness within environmental psychology of the interrelatedness of object and subject' (Whitehead 1989, 258). Softer, less hard-line non-laboratory research methods are already being used with success (R. Kaplan and S. Kaplan 1989, Porteous 1991).

Most of the work in experimentalist environmental aesthetics investigates the 'preferred scenes' of subjects. The assumption that preferred scenes are somehow good for us, with the corollary that disliked ones have the opposite effect, has rarely been tested. Yet it would seem especially important to discover if scenes that are 'good to see' are simply good aesthetically (i.e. good to think or good to feel) or whether such aesthetic satisfactions have deeper implications, such as the promotion of physical well-being and mental health.

Urban versus rural

Despite the urbanist orientations reported by Phillips and Semple (1978) on the part of planning students, a generation of work in experimentalist environmental aesthetics has repeatedly confirmed that respondents in laboratory settings overwhelmingly prefer natural (rural or wilderness) over urban scenes. This is perhaps the most significant finding of the experimentalists. It has been confirmed by over one hundred studies, not merely in the Western world (Ulrich 1976, 1993) but also in more traditional societies (Hull and Revell 1989). In view of the traditional anti-urban bias in the Western world, the direction of the preference is less surprising than its strength.

In Nasar's collection of papers entitled *Environmental Aesthetics* (1988), the 300-page central section reports empirical research by experimentalists on both architectural interiors and exteriors, and an array of urban and nonurban scenes. The nub of the findings is: that environmental evaluation involves three major components, pleasantness, excitement, and distress; that the salient perceptual and cognitive features in promoting pleasantness and excitement are visual richness (complexity, mystery), clarity (legibility, coherence) and the unobtrusiveness of built elements. In terms of visual preferences, there is great variety according to observers, scenes and kind of activity studied. In general, however, aesthetic quality expressed as preference appears to be enhanced by coherence, order, complexity, and compatibility of scene elements, and by the appearance of naturalism. Aesthetic quality is almost universally reduced, especially in nonurban scenes, by manmade intrusions.

A classic study by Kaplan, Kaplan and Wendt (1972) illustrates the point. Subjects were shown slides arranged along a 'nature–urban continuum' broken into four sets: urban (stores, traffic, factories, tall buildings, etc.); more urban than natural (mixtures of suburbs, intersections, apartments set in grass and trees); more natural than urban (a similar mix with a different emphasis); and 'nature' (trees, meadows, etc.). The rather spectacular results were:

> Nature material was so vastly preferred over the urban slides ... that the distributions [of preference scores] barely overlap. In other words,

with a single exception, the least preferred nature slide was favored over the most preferred urban slide (p. 355).

With minor variations (R. Kaplan 1973a), a host of subsequent studies have confirmed this pro-nature bias (Figure 3.8).

A series of studies on roadside aesthetics, for example, discovered that open forest (the savannah again) was more preferred than dense forest, which was favoured over flat farmland. If the latter was manipulated by planting a few trees, however, its perceived aesthetic quality was likely to rise (R. Kaplan 1977a). Evans and Wood (1980) used slides of roadsides undergoing 'sympathetic' versus 'unsympathetic' development, for example agriculture and rustic fences versus housing and metal fences, and found that 'even slight changes in adjacent roadside development effect significant changes in perception of roadside quality' (p. 268). With increasing development, respondents felt that the landscape progressively became more worthless, useless, cluttered, unpleasant, ugly, and drab. In a reversal of the usual procedure, Winkel (1973) found steady increases in the perceived quality of scenes as poles, wires, billboards and other tokens of urban clutter were systematically deleted from photographs.

Even when quite unspectacular nonurban scenes, such as ordinary roadside verges or undistinguished farmland, are shown to experimental subjects, nearly always the least-preferred nonurban scene – whether of wilderness or countryside – is scored higher than the most-preferred urban scene. It is as if we are dealing with two environments that are so differentially judged that they can hardly be scored on the same scale. And

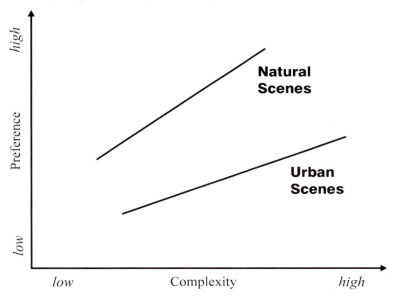

Figure 3.8 Natural and urban scenes judged in terms of preference and complexity

in urban scenes, the greater the quantity of vegetation, the higher the rank. Nothing could better confirm the traditionalist city planner's emphasis on the importance of urban parkland.

The uses of scenery

What do we wish to do with or in these wilderness, rural, or leafy urban scenes? According to Ulrich and Addoms (1981) we chiefly wish to contemplate them. In this study several groups differing widely in social and recreational activity levels tended to agree that the most important value of urban parkland was 'the passive contemplation of nature'. Even non-users of parks felt better knowing that the park existed, an argument favouring the establishment of remote national parks and ecological reserves which are little visited by the public.

Two studies on gardening carried the argument a stage further. R. Kaplan (1978b) surveyed over 4,000 members of the American Horticulture Society; over 60 per cent of these largely middle-class persons cited feelings of 'peacefulness and tranquillity' as the most important satisfaction to be obtained from gardening. Lewis (1979), in contrast, surveyed low-income urban gardeners working plots in public-housing projects, vacant lots, and tenement rooftops in New York, Philadelphia, Chicago, and Vancouver. Collectively, an active urban-gardening programme was found to promote sociability, reduce vandalism, and generate neighbourhood revitalization, including cleaner streets and building paint-up. For the individual, self-esteem was considerably enhanced and there was a general improvement in 'a sense of tranquillity and well-being'. Physicians in a senior citizens' centre reported more objective measures: 'It helped lower ... blood pressure; some patients are taking less medication, are more relaxed, and feel they are needed' (p. 333). Not surprisingly, urban-gardening projects have been strongly promoted by such agencies as the United States Department of Agriculture and the British Columbia Housing Management Commission.

In a world of concrete, steel, and stone, the urban garden or park interfaces between raw nature and the world of technology. Both urban gardening and the passive contemplation of vegetation appear to have a healing quality. Plants are long-term, not immediate, satisfactions; they are steady in growth, not erratic; they stay in place; they are quiet rather than noisy; their rhythms are slow and natural, not the up-tempo metropolitan rhythms of future-shock. Experimentalist research, then, confirms long-held beliefs about the life-enhancing value of contact with nature.

If nature is so valuable to us, will we go out of our way to make contact with it, even when such contact is economically inefficient? Urban trans-portation planning is generally premised on the myth of Homo economicus, who demands the fastest route between origin and destination. We therefore assume that in a journey, say, from home to work or a shopping centre, the car driver will choose the perceived shortest distance between the two land uses.

To test this proposition, Ulrich (1974) devised a field survey which involved automobile drivers choosing to take ordinary, functional trips to shop on either a non-scenic, shorter, expressway (with permissible speeds up to 70 mph) or a longer, scenic parkway (35 mph). Even though all subjects knew that the other route was faster, 54 per cent of all trips were made on the parkway; indeed, only 4 per cent of the respondents conformed to the Homo economicus model in making all their trips on the fastest route. People in general like variety, women in particular found the expressway dangerous, and men valued the expressway chiefly as an opportunity to drive fast (partly an aesthetic sensation!). Users of the parkway, however, stressed that non-economic considerations were important to them in choosing the slower route. The most important non-economic benefit reported was the scenic visual environment of the slower route.

Further, all drivers, whether or not they habitually used the parkway, agreed substantially on the elements which contributed to the higher visual quality of the scenic route. Not surprisingly, visual complexity and the pleasingness of a park-like landscape proved most important. As both parkway and expressway users had similar landscape tastes, their differences in route choice reflected the differential value they placed on scenery. It is clear, moreover, that more people place a high value on scenery, even when on functional trips, than engineering-based road-systems planners would have us believe. Intuitive road design may thus be as important as the prevailing time-efficiency model.

The shopping-trip study took us from studies of passive aesthetic preference to the study of active aesthetic behaviour. Pressing yet further, Ulrich then sought to test the humanist intuition that environments we regard as pleasant have a direct effect on psychological and physiological well-being.

The 'nature tranquillity hypothesis', very ancient in origin, holds that the contemplation of nature is psychologically healthful. Contact with water and plants, especially, will reduce stress, promote relaxation, and induce good feelings (Herzog and Bosley 1992). The notion has long been basic in urban design and was explicitly stated as a major planning principle by planners as influential as Frederick Law Olmstead.

Ulrich (1979b) tested the hypothesis by using subjects in a high state of anxiety arousal (students shortly after an examination!). Subjects first answered the Zuckerman Inventory of Personal Reactions (ZIPERS) and were then asked to rate colour slides of scenery on an aesthetic preference scale. Finally, a second round of ZIPERS was undergone.

Subjects were divided into two groups, one of which saw 'ordinary natural scenes' (e.g. rural settings), the other urban scenes with no nature content. Before the screenings, both groups tested high in anxiety, and were especially high in anger and fear arousal, and correspondingly low in positive affect. After the slides, the viewers of natural scenes showed an increase in positive affect and decreases in fear arousal, sadness, and anger. In sharp contrast, the viewers of urban scenes tested even lower than before on positive affect, and while fear arousal decreased slightly, both sadness and anger increased.

This was an extremely conservative test, for the nature scenes chosen were often scruffy rural views, while the urban slides depicted not extremely ugly scenes but relatively unblighted environments. The aetiology of such effects is unknown; we may be dealing with a strong cultural bias against cities which has become an archetype. It can be stated with confidence, however, that the intuitions of the ancients and the recent research on parks and gardening are substantially correct; exposure to nature is therapeutic.

Delving yet further, Ulrich (1981) sought to discover whether this expressed preference for therapeutic nature over anxiety-laden urban environments might have some physiological basis. In the next experiment subjects were exposed to a (surprisingly incomplete) array of slides in three categories: nature dominated by water; nature dominated by vegetation; and urban scenes with neither water nor vegetation. The subjects were monitored in three ways: brainwave alpha-amplitude levels, using an electroencephalograph (EEG); heart-rate levels, using an electrocardiograph (EKG); and emotional state, using ZIPERS and semantic scale tests.

No heart-rate variations among groups appeared on the EKG. The EEG, however, recorded significantly higher anxiety (lower alpha-wave amplitude) on the part of subjects viewing urban scenes. Vegetation reduced anxiety somewhat, but water significantly lowered anxiety levels. As expected, the tests of emotional states found urban scenes to heighten, and nonurban to lower, feelings of sadness and fear arousal. Again, vegetation lowered fear arousal somewhat, but water scenes were associated with a marked reduction in fearfulness.

Figure 3.9 The nature tranquillity hypothesis: everyday therapy in Victoria, BC

Figure 3.10 Here water is explicitly sacred: Varanasi, India (courtesy Gavin Porteous)

Clearly, there are physiological as well as purely cognitive reactions to the scenic qualities of environment. The soothing and relaxing effect of water is notably confirmed. Those who sit in cars and on benches and stare at the ocean are, as we knew all along, doing themselves a power of good (Figures 3.9, 3.10).

Finally reaching the real world, Ulrich (1984) abandoned the laboratory for a hospital setting. In a well-designed, highly-controlled study, matched groups of patients in a state of high anxiety after cholecystectomy surgery were placed either in a room with a view of trees or in an identical room with a view of a brick wall. Not surprisingly, in view of Ulrich's earlier sequence of experiments, the 'tree-people' fared better than those whose view was of inanimate brick. Specifically, tree-people enjoyed shorter post-operative stays in hospital, took fewer potent analgesics, tended to have fewer complications, and earned fewer negative comments in nurses' notes, thus implying more positive attitudes or easier recovery. The work has been replicated (Ulrich 1993). The implications of such findings for hospital design, management, and cost-efficiency are great. For the individual patient, pleasanter environments and faster, cheaper recovery may result. On the grand scale, the implications for a future theory of environmental aesthetics involving both psychological and physiological components, married perhaps to an Appleton/Kaplan evolutionary thesis, are exciting and profound.

Parsons (1991), in an exhaustive survey, provides further support for the evolutionary theory and for Ulrich's approach to research on the restorative value of aesthetics. First, he reviews behavioural evidence which suggests that human emotional response is, at least in part, evolutionarily driven. Second, neuropsychological evidence suggests that the extremely quick, gross processing of incoming stimuli, as would be expected of an evolutionary capacity, has clear survival value. Because this processing of stimuli is completely subcortical, centring in the limbic system of the brain (see Chapter 1.3 above), it is likely to be affective and result in preferences. Third, research on stress-related immunosuppression mechanisms suggests complex endocrine response systems as the pathways whereby sensory contact with nature results in anxiety reduction.

In sum (Parsons 1991: 1):

> Research in the field of environmental perception has often shown that people in a number of different cultures have similar visual preferences for natural environments over urban environments. Interestingly, these preferences are consistent with the old and widely-held belief that exposure to nature is salubrious ... [S]tudies have focused on the anxiety-reducing effects of visual exposure to natural environments ... indicating that immediate affective responses to environments may influence environmental preferences (as well as the other subsequent emotions) and trigger physiological processes that can influence the immune system, and thereby, physical well-being.

People continue to flee the city for recreation. City-related stress continues to cause a wide array of serious illnesses (Ulrich 1993). In these circumstances, the least experimentalist aestheticians could do is to mount a massive research investigation of the physiological health consequences of 'natural' stress-reduction.

3.4 METHODOLOGICAL PROBLEMS

It was once believed that the very deep, private feelings involved in death, aesthetics, religion, and sexuality might never be amenable to the experimentalist methods of laboratory, survey or questionnaire. Back and Bourque (1970), however, assert that meaningful work can be done in these areas, and that the results are quantifiable; 'This opens up the possibility of studying questions of emotions as social phenomena, which would be impossible if their study were restricted to the special conditions of the psychiatric depth interview.'

In environmental aesthetics, however, little attempt has been made to delve deeply into the emotions. As yet, most work remains near the surface, dealing primarily with issues such as preference.

Most preference and related studies follow a similar procedure, outlined by Wohlwill (1976) as:

1. select and measure the environmental variables to be studied (e.g. complexity, texture);
2. sample a range of environments (E) to provide an array of environmental variables (e.g. urban, nature, etc.);
3. sample research subjects (Ss) for an array of personal variables (e.g. age, sex, etc.);
4. present E to Ss, usually in simulated form such as colour slides;
5. record Ss' responses on a variety of response measures, and analyse and interpret the responses.

Each stage of the procedure has its problems.

Selecting environmental variables

Much early work in environmental aesthetics simply failed to measure the chosen environmental variables in any meaningful way. Scenes might be roughly classified by the researcher into 'moderately complex', 'very complex' and similar categories, but measurable assessments of the degree of complexity were not made. More recently, researchers have resorted to panels of judges, often experts in the field, who are asked to sort a mass of slides into a predetermined number of categories based on a particular environmental variable. Only sets which show a very high level of agreement among judges should be used. Wohlwill (1976), however, notes that confusion arises between actual attributes of environment and judges' affective responses (he cites early Zube work here), and when assessment judges are subsequently used to generate evaluative responses (early Kaplan work).

Some variables can be assessed rather more objectively. Fabos *et al.* (1975) have developed measures of land-use diversity (number of land uses and proportion of area in each) and land-use contrast (e.g. texture, edges between uses, etc.), while methods used by geographers, cartographers, and landscape architects are also valuable (Kaplan 1977a, b). Finally, multidimensional scaling procedures can be used to describe natural landscape configurations (Shafer 1969). The problem remains whether such measures are accurate reflections of what observers actually attend to when viewing environments.

Sampling environments

Most environmental psychologists producing slides or photographs to present to subjects have learned to control as many environmental variables as possible, notably sun angle, temperature, vantage point, sky content, and so on. Seasonality, for example, can seriously affect responses (Buhyoff and Wellman 1979).

Theoretically, one should sample from every one of a large number of small grid units, taking views in every possible direction! In practice, researchers confine themselves to a small number of localities and control for views from certain angles only. Thus the extremely broad range of conditions which confront an observer in the field is never accurately represented by a set of surrogate slides or even movies. Much more work is required on this issue.

Sampling subjects

Over a century of serious psychological experimentation has ensured that the sampling of subjects has been refined almost to perfection. The problems in much psychological research, however, relate to the subject pools from which respondents are drawn. When not working with rats and other non-humans, psychologists typically draw their subjects from 'subjected' groups, notably undergraduate students and low-ranking members of the armed forces. The problem, then, is the generalizability of results from such restricted groups. One cautiously notes a body of studies (Winkel *et al.* 1970, Zube, Brush and Fabos 1975, Evans and Wood 1980) which suggests substantial agreement between students and representative samples of the public at large. Further, Zube contends that personal landscape assessments are relatively fixed by adolescence, so that students should have similar responses to their older counterparts.

Presentation of E to Ss

This stage of the research procedure has generated most controversy. Practically, most of this work has taken place under laboratory or quasi-experimental conditions. In consequence, little experimental work has involved subjects' assessments of real-world environments. In general, the environmental displays presented as stimuli involve some form of simulation. Such simulations may be static (a photograph) or dynamic (a film), abstract (a computer model) or concrete (a scale model) (Craik 1970). Most research has used concrete, static displays, especially colour slides and photographs.

Validation studies suggest that although observers' reactions to colour photographs are moderately predictive of their on-site responses (Seaton and Collins 1972), many other displays are of less value, and none compare with using the real environment as stimulus. Some attempt has therefore been made to use dynamic simulations. Films and videotapes have been brought into play, thus providing some of the environmental attributes absent from static displays. Indeed, Wohlwill (1978) suggests that it may be the soundless, movementless nature of slides that accounts for most of the differences found between responses to slides and actual scenes. The same attributes may also contribute to the well-known preference for rural or natural over urban slides. According to Wohlwill, a photographic simulation of natural scenery captures more of the total environment of such a scene than does a photograph of an urban area, the ambience of which is far more dependent upon man-made sonic and dynamic components.

Although an extremely wide variety of environmental simulation displays is possible (Craik 1970), simplicity, economy, and convenience generally dictate the use of colour slides. Seaton and Collins (1972) found these to be the best of a wide range of surrogates, a finding confirmed by Danford and Willems (1975) with the proviso that all studies use strict controls to check possible instrument bias. Shuttleworth (1980) found colour photographs to be valid surrogates for real landscapes as long as: the sample of environments was large, and sampled from a variety of landscape types; the sample of respondents was also large, and representative of all sections of the community; both the real landscape and the simulation were judged by respondents; both landscape attribute assessments and preference reactions were elicited; and all photographs were taken with a wide-angle lens to accommodate lateral and foreground context without the distortion of the actual scale relationships that are found in direct observation of the landscape.

Coeterier (1983) performed a validation study of Shuttleworth's criteria and reported that colour photographs were a good surrogate for reality when depicting small-scale landscapes and when judging such characteristics as 'intensity of human use' and 'historical character'. Photographs, however, are poor substitutes for the actual landscapes in the case of large-scale landscapes with great depth of field; they also miss microrelief and small terrain variations.

Colour slides also performed fairly well in a study of the effectiveness of architectural displays (Acking and Kuller 1973). The experimenters reported that coloured models most accurately mimicked real architecture for the respondents, though colour slides of the models taken from eye-level were almost as good. Acking and Kuller concluded by taking architects to task for continuing to represent their ideas in far less intelligible forms such as plans, perspective drawings, and white schematic models. Large-scale models such as the Berkeley Environmental Simulator are the presentation medium of choice (Chapter 5.2). Given the cost of such instruments, it is clear that colour slides will remain a major surrogate for real environments in the near experimentalist future.

Response measures and analysis

The measurement of respondents' perceptions or assessments has generally been achieved through verbal rating scales such as the semantic differential. These are usually analysed through intercorrelations between ratings, which are further reducible by factor analysis to a small number of dimensions. The problem here is that the result is purely descriptive; we gain little knowledge of the effects on affect of specific environmental characteristics. For instance, although it is common to find that nonurban scenes are usually rated as much more pleasant than urban ones, little information is available to specify which attributes, such as complexity, are salient in this assessment. As Wohlwill (1976: 68) states, we learn little about 'functional relationships among variables, which, one would suppose, is the business of a scientifically-based research effort'. Both Evans and Zube (1975) and Garling (1976) have also severely criticized the general reliance on semantic differential procedures and suggested that alternatives might be derived from the promptings of theory. The general tendency, however, has been to develop ever more complex batteries of procedures (Pearce and Waters 1983).

Clearly, while great attention has been given to technique development and the generation of empirically-derived models of preference, too little attention has been paid to the development of related theory (Zube, Sell and Taylor 1982). Wohlwill's extension of Berlyne's theory of aesthetics and the Kaplans' (1989) landscape preference prediction work are exceptions to this trend. Meanwhile, the less theoretical have expanded their range of assessment techniques to include a wide array of perceptual and behavioural techniques, including psychophysical scaling (Ulrich 1979b), behavioural mapping (Bechtel 1967), and physiological monitoring procedures (Ulrich 1981). Stokols (1978) suggests that little information exists on the relative validity of this battery of simulation and measurement procedures.

Finally, the environmental philosopher Carlson (1979) takes experimentalists to task for their almost exclusive reliance on visual simulations. He believes that the use of such stimuli overemphasizes the formal qualities of environment, such as the unified/chaotic, balanced/unbalanced dichotomies which occur so often in experimentalist work, while neglecting the sensuous qualities, such as texture, colour, smoothness, and the like. For Carlson, the use of photographs is a formalist way of looking at the world, rather like the eighteenth century's Claude glass.

Carlson goes on to argue that the formal properties of environments are unduly emphasized in photographs, which presuppose both a framed object and an external observer who is separated from the object, viewing it passively from an external viewpoint. He points out that observers in real landscapes are inside the frameless scene, and are actively involved with it in a multisensory manner. In other words, experimentalist research is gauging subjects' reactions to objects which are more like formal landscape paintings than real environments.

It is clear that tourism management has begun to create 'scenes' out of 'environment', as when the tourist stops at a scenic viewpoint and frames a view in the camera. These views are often repeats of classic views taken by nineteenth-century photographers. Thus nature imitates art. Nevertheless, Carlson's point weakens the ground beneath most modern experimentalist research, and deserves a more considered response.

Beyond the laboratory

By the mid-1980s it had become apparent that the traditional, hard-science, laboratory-based experimentalist model was in many ways inadequate to the task of understanding aesthetic person–environment relationships. One basic problem has always been the 'conspicuous theoretical void' in experimentalist environmental aesthetics (Zube, Sell and Taylor 1982: 25). Priestly (1983) confirmed that in the 1970s there was no evidence of a convergence toward a unified theory of landscape perception.

Sancar (1985) has attempted to provide such a theory. Its basic assumption is the centrality of 'the interaction between humans and landscapes', and thus 'any process of inquiry in landscape perception should allow for the participation of the observers in purposeful activities relevant to the landscape in question' (p. 119). In other words, unless respondents are studied as they pursue their normal activities in real situations, rather than performing unusual tasks in laboratory settings, experimentalist aesthetics has no justification for its attempts to influence public landscape policy (see Chapter 5, below). Sancar locates various research paradigms on active–passive and holistic–dimensional axes (Figure 3.11); if my four basic approaches (Chapter 1.1 below) are fitted on a diagram derived from Sancar, it is clear that current experimentalist work is far too passive and dimensional, and should make tracks forthwith to occupy a position, close to that of planning, which is more active and holistic.

Such a position requires a new research paradigm to replace the old either/ or, environmental attributes versus personal determinants, emphases. This new 'integrative approach', involving a sequence of identifying actors, identifying situations, information pick-up, action, and feedback, is explained in Sancar (1985). Similarly, Fenton and Reser (1988) reject the traditional division of effort into 'objective' and 'perceptual' approaches and demand an 'interactional perspective'. They affirm that 'the perception of landscape quality is an interactional phenomenon ... simultaneously dependent on both objectively definable landscape variables and an individual's knowledge or cognitive representation of the landscape. There is an urgent need for an integrative theoretical and methodological approach to environmental perception, which would recognize that it is the perceived as well as the real world to which we respond' (p. 110). In a similar vein, Whitehead (1989) asserts that there is now in environmental psychology a 'very widespread awareness ... of the interrelatedness of object and subject'. If common sense were not at a discount in the academic world, this approach would be called common sense.

A major body of substantive work in this 'interactionist' or 'integrative'

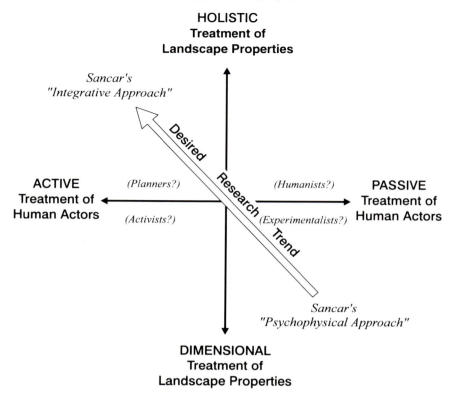

HOLISTIC
Treatment of
Landscape Properties

Sancar's
"Integrative Approach"

Desired Research Trend

ACTIVE
Treatment of
Human Actors

(Planners?)

(Activists?)

(Humanists?)

(Experimentalists?)

PASSIVE
Treatment of
Human Actors

Sancar's
"Psychophysical Approach"

DIMENSIONAL
Treatment of
Landscape Properties

Figure 3.11 Alternative approaches to research (based on an idea in Sancar 1985)

mode has yet to appear. Of significance is the Kaplans' bold decision to shed their psychologists' white lab coats (so necessary to self-esteem in a discipline whose goal is to become the physics of the social sciences) in favour of naturalistic studies in real-world settings. In their chapter 'A Wilderness Laboratory' (Kaplan and Kaplan 1989) they are nevertheless at pains to demonstrate that the 'soft' methods of field observation, journal-keeping, depth-interviews and the like, are respectably scientific in the sense of involving 'a process of guided discovery, where theory is refined by data and data gathering is motivated by theory' (p. 121).

Their study of participants in the Outdoor Challenge Program in Michigan's Upper Peninsula comes very close to the work done by geographers on recreation in wilderness settings. In particular, their discovery that wilderness experiences often produce feelings of 'wholeness' or 'oneness' confirm traditional humanist views of nature as a source of spiritual renewal. These quasi-religious feelings, as described by both Thoreau and modern Outdoor Challenge teenagers, proved surprising to the Kaplans: 'A surprising outcome of the wilderness research has been the remarkable depth of such feelings' (p. 147).

The results are far less surprising to humanist geographers who have engaged in field studies coupled with close reading of published experiences of wilderness travel. My own research in this area involved a study of transcendental experience in wilderness (Porteous 1991).

Transcendental experience is achieved when one attains a level of felicity or exaltation such that one is 'taken out of oneself'; the self is transcended, and a higher state of consciousness is achieved. This ecstasy (literally 'standing outside oneself') may be achieved in many ways, including the geographical, notably in the experience of exotic landscapes, close encounters with nature, hard travelling, and situations involving remoteness, isolation and desolation. When this rapture involves a sudden opening-up to meta-physical forces in a wilderness setting, it has been termed a 'theophany' (Graber 1976).

Although they are frequently reported by poets and painters, the question of whether transcendental experiences occur to less expressive members of the public in wilderness settings has received little attention (Kaplan and Talbot 1983). In August 1989, therefore, I took part in a sea-kayaking expedition to Ellesmere Island, at 80° North in the Canadian High Arctic. The uninhabited landscape proved severe, fjords were full of icebergs and floes, air temperatures rarely rose above 2°C, daylight was continuous, many huge sea mammals were encountered, and snow fell on 15 August (Figure 3.12). Data were collected via participant observation, journal-keeping, and extensive post-expedition in-depth interviews.

Figure 3.12 Author and crew engaged in experiential research in the Canadian Arctic, August 1989

Three kinds of intense experience characterized the expeditionaries. First, most felt more 'alive' than normal, especially in situations of danger not usually encountered. Improved respect for natural forces resulted, in conjunction with, almost paradoxically, improved self-esteem. This confirms Burton's (1981) exhaustive study of 300 wilderness programmes, which found that the most frequent outcome of such trips is improved 'self-concept'. Second, encounters with animals and icebergs provoked the largest number of expressions of awe and aesthetic beauty. Some were entranced by the icebergs' 'unbelievable blue, like no colour I've ever seen, a glowing, radiating colour, very close to a spiritual feeling'.

Deeper, more truly transcendental experiences, however, were enjoyed by only a minority, and had little to do with physical exertion, danger, or animal encounters. Rather, these feelings were most likely to be aroused by being alone and still in the landscape. A sense of the unerring rightness of the physical universe flowing through time, a calm acceptance of one's insignif-icance, and a feeling of deep harmony between self and world were the essence of this transcendental experience of nature, Wordsworth's 'sense sublime/Of something far more deeply interfused'. As one kayaker put it:

> having escaped from the constraints of myself, I was able to appreciate the world as if for the first time ... I had a sudden feeling of rightness, a sense of everything fitting together, a feeling of one-ness to the point that I did not want the moment to stop.

It is unlikely that the transcendental experience of nature can be deliberately induced. It is possible, however, to specify some conditions in which it is more likely to occur. These include: being alone, or with a trusted companion; being passive and receptive, as late at night after an exhausting day; and being silent in the presence of a vast and powerful landscape. Danger, physical exertion, and sudden encounters with animals may result in an extremely high level of exhilaration, but for most there are chiefly feelings of environmental challenge, self-encounter, and enhanced self-esteem in such experiences. On the contrary, transcendental experience appears to involve environmental humility, a childlike acceptance of what is, and a feeling of wholeness, harmony, and at-one-ness with the landscape. Its essence is not challenging, but waiting; not doing, but being.

It is interesting that the three components just outlined closely resemble three of the four axes of my concept of Being (Figure 1.3). Attachment, of course, cannot be expected of visitors spending only a brief time in an area.

What is clear from these tentative theoretical discussions and pioneer research efforts is that the experimentalist future will be far less experimen-talist in the narrow laboratory sense of the term. Until now the paradigm has been characterized by unresolved theoretical and methodological issues, a splintered research effort involving either stimulus-defined or response-based approaches, a gaggle of miscommunicating disciplinary orientations, and a general failure to 'develop an approach that would allow for a meaningful and working relationship between theory and method' (Fenton and Reser 1988: 108). Such a relationship is not merely a consummation devoutly to be wished (Wohlwill 1976, Carlson 1977, Zube, Sell and Taylor 1982). It is the critical, bedrock prerequisite for any further meaningful development of experimentalist environmental aesthetics.

The value of experimentalism

Nineteenth-century conceptions of aesthetics were bold and flamboyant. One thinks of the fascinating speculations of Francis Galton, William James, and the sensualist radicals engaged in the 'aesthetic adventure' (Gaunt 1988). Late twentieth-century experimentalist environmental aesthetics is far more timid, for its model is psychology, which in turn suffers severely from the condition of physics envy.

The hard-science model has, nevertheless, certain advantages. It is, or should be, extremely self-critical; hence the preoccupation with method-ology. Through the teasing-out of the relationships between environmental attributes, personal characteristics, and affect or response, the experimen-talist approach permits the operationalization of aesthetic concepts, partic-ularly in terms of constructs such as preference.

Perhaps the most exciting aspect of experimentalist environmental aes-thetics is the support it provides for a fledgling evolutionary and physio-logical theory of the origin of the aesthetic impulse. Here experimentalist and humanist approaches confirm and extend each other, providing some hope that this largely atheoretical interdiscipline will one day develop a convincing theoretical apparatus.

Experimentalist work has also been extremely fertile in confirming ancient humanist concepts such as the 'nature tranquillity hypothesis' and the importance of attributes such as complexity and mystery in the generation of affect. It is easy to dismiss such work as merely confirming what we all know already (Ackerman 1990) but given a socio-political environment which demands scientific studies, quantitative data, and evaluations as a matter of routine, it is valuable to generate hard data to support what is intuitively known and unreflectively practised.

Experimentalist work is thus extremely valuable in forming a bridge between humanist intuition and the realm of action. Research on the positive relationship between nature and personal well-being may well form ground-work for the generation of design and policy in such areas as psychotherapy, architecture, park planning, and hospital management. In this respect, experimentalists have been extremely active in the field of landscape assessment for planning, which is reported below in Chapter 5.

The experimentalist position necessarily involves an attempt to adopt a value-neutral attitude towards environmental stimuli. This, to some extent, also characterizes the more scientific aspects of planning. Activists, however, are anything but value-neutral, as I will demonstrate in the following chapter.

4

ACTIVISTS

Logging truck

O if we but knew what we do
When we delve or hew –
Hack and rack the growing green!
Since country is so tender
        ... even where we mean
To mend her we end her,
When we hew or delve:
After-comers cannot guess the beauty been.
                                Gerard Manley Hopkins

Activists are essentially people with values who are ready to act upon them. In sharp contrast, both humanists and experimentalists are generally quietist observers, the latter with a sense of detachment which comes from vain attempts to engage in value-free research.

Values are the basis of action. Activists derive their values chiefly from a humanist cultural context (Chapter 2). I doubt that many are fully aware of the extent to which experimentalists (Chapter 3) have begun to confirm traditional humanist beliefs. Activists are conservative radicals, with a strong thrust toward preservation, conservation, and heritage values. They question the material progress ethic which underpins both capitalism and communism, and they are often 'green', in more than one sense of the word.

Most radical movements in environmental aesthetics begin with an elite, move to mass public support, and, if fortunate, end in elite action at the institutional level. In other words, protest against or protest on behalf of, on the part of small groups (Chapter 4.1: Literary and design activists, below) is frequently followed by widespread support (4.2: Citizen action), legal action (4.3), and the creation of public policy (4.4). Most activists agree that, in the long term, their needs will be served best by environmental education (Chapter 5.3).

Activism varies from the small-scale (protests against the destruction of attractive houses or city trees by developers) to the large-scale (campaigns to save tropical rainforests or develop national aesthetic standards for roadside development), from the short-term (temporary union 'green bans' on dubious development projects) to long-term (environmental education in elementary schools), from the relatively ineffective (letters to the editor) to the more effective (legislation), and from the illegal (ecotage) to the almost respectable (forming a green political party). Above all, activism stirs the minds of planners (Chapter 5), politicians, private business people, and the public, and gives them the opportunity to reconsider their positions *vis-à-vis* their aesthetic and ethical relationships with their surroundings.

4.1 LITERARY AND DESIGN ACTIVISM

Protests by literary and design professionals against low aesthetic standards in the environment have been common since Roman times. During certain periods, most notably the seventeenth and eighteenth centuries in Europe, active aesthetes were able to reshape whole countrysides and townscapes to conform with contemporary environmental aesthetic standards. The Industrial Revolution bade goodbye to all that, ushering in a crass cash-nexus and the primacy of economic attitudes, the environmental results of which are with us today, and especially apparent in countries which have undergone most of their landscape transformations after the eighteenth century.

Literary activists

Poets and novelists may take on the role of a nation's conscience. They are certainly far more attuned to aesthetics than their contemporaries, and are thus invaluable as early-warning devices for impending aesthetic problems. Since the late eighteenth century a myriad social novelists have drawn our attention to the growing ugliness associated with urban industrial life. Even the most introspective, such as Malcolm Lowry, display a keen environmental aesthetic sensibility (Porteous 1992). Yet, from Horace, who railed at the ugliness of Rome, to Wendell Berry and Bill McKibben (1989), whose poetry and polemics flay current American lifestyles, poets and novelists are not well placed to get their message across to the public. By the late twentieth century, however, it is notable that ecological and environmental aesthetic concerns are occasionally aired even on television soap operas, which clearly reach a much greater audience today than do poets and serious novelists.

Nevertheless, the example of two poets active in late twentieth-century Britain stresses the importance of poetic authority in shaping the public mind. Britain retains the official position of Poet Laureate, occupied until his death in 1984 by Sir John Betjeman. He should have been succeeded, in the opinion of many, by Philip Larkin (died 1985). The point about these poets is that their work was easy to read and thus had a very wide audience.

Philip Larkin's voice was 'one of the means by which his country recognized itself' (Motion 1993: 343). According to the writer of a letter to the editor of the London *Guardian* (Healey 1989), Larkin was above all concerned with the unvarnished truth: 'the truth about one's relationships with one's parents' ('They Fuck You Up, Your Mum and Dad'), the truth about modern life ('High Windows'), the truth about childhood and growing up ('I Remember, I Remember'), the truth about home ('Home Is So Sad') and the truth about being old ('The Old Fools'). His posthumous *Collected Poems* (Larkin 1988) sold 35,000 copies in the first two months after publication, putting it among the British hardback best sellers, whence it shone out from among the usual titles in such lists with a hard, gem-like flame.

Larkin celebrated the aesthetic beauty of very ordinary landscapes, such as the townscape of the dull, working-class city of Hull, East Yorkshire, a place of 'domes and cranes' with a skyline of 'wharves and wires, ricks and refineries', where the tall new hospital shines like a 'lucent comb' amid 'a fishy-smelling pastoral' (Spooner 1992). He comments wryly on both modern lifestyles and townscape in his poem 'Essential Beauty', which describes huge urban advertising billboards in all their incongruity and shows how our quest to live up to the impossible standards set by such advertising blights both our lives and our landscapes. 'Going, Going', Larkin's most famous and telling environmental poem, was commissioned by the British Ministry of the Environment. It captures exactly the sad, mild, impotent environmentalist feelings held by the British silent majority, middle-aged and middle-class, a belief that all that is pleasing in the English landscape will be swept away by highways, bleak high-rises, and rampant profit-oriented development.

If Larkin's role was to deplore, in sadness, the despoliation of England, Betjeman's was to celebrate ordinariness, point out the vulgarity and aesthetic vandalism of modern life, and ridicule its perpetrators, whom he conceived, rather narrowly, to be bureaucrats and planners. Betjeman was known as the poet of the suburbs, where most people now live, and had a deeply sympathetic insight into the minds of suburbanites; 'he stands for the small, the local, the kindly' (Birkenhead 1980, xxv), an ethos appealing strongly to the English mentality.

His anti-planning poems (Betjeman 1980) are violent but cheerful. Seeing the vast red-brick ugliness known as the town of Slough, he prays:

> Come, friendly bombs, and fall on Slough
> It isn't fit for humans now . . .

Out in what remains of the periurban countryside he pens a 'Harvest Hymn', wherein farmers spray poison on their fields to oust 'wicked wildflowers', and laments the loss of country lanes to the ever-increasing needs of speeding motor cars. In an urban nation with a positive ethos of country living, these shots go directly home.

Planners and architects, with their visionary schemes of engineered perfection, become Betjeman's especial target. In 'The Planster's Vision' we find an urban designer cutting down trees and destroying country cottages in favour of a future which brings together in one image the bleak exurban landscapes of both capitalism and communism, where:

> . . . workers' flats in fields of soya beans
> Tower up like silver pencils, score on score . . .

For those brought up on Orwell's *1984* and in the 'communal canteens' of war-time and postwar Britain, these verses were telling. Unfortunately, although planners are today less visionary, they are still judged responsible for a great deal of landscape desecration. Betjeman appeared to have no conception of the corporate forces that drive planners, politicians and bureaucrats, but was aware of the aesthetic destructiveness of small-scale enterprise which makes people protest, though vainly.

Finally, as with Larkin, no programme for the alleviation of aesthetic landscape misery appears. It is not the duty of poets to set forth plans of action. Their activism consists in celebrating aesthetic values and pointing out the doleful effects of the modern lack of aesthetic interest in landscape.

Design activists

Humanists are contemplatives; their role is to analyse culture and to generate ideas. There is, however, a considerable body of humanist landscape critique which, though not strictly activist, does provide a solid critical base for action. Humanist landscape critique most commonly deals with urban visual blight (Lewis *et al.* 1973) and the growth of placelessness.

In *Place and Placelessness* (Relph 1976) discusses first the concept of place (meaningful space) as opposed to space, which is amorphous and contextual. In *Space and Place* (1977), Tuan makes the same distinction. The essence of places includes notions of location, landscape, time, community, privacy, home, identity and authenticity. Places are, above all, different from each other, and this provides the mobile with a continually varied aesthetic experience.

Relph's point is that modernizing tendencies in both urban and rural areas are rapidly expunging place in favour of space. Mass communications, mass culture, mass tourism, big business, big labour, commercialization, commodification, and centralizing bureaucratic tendencies throughout the world are all synergistically engaged in reducing the diversity and significance of places through the processes of homogenization, standardization, simplification, and disneyfication.

Two generations ago, J.B. Priestley (1955) roamed an American landscape that was already beginning to homogenize; to contrast it with the Soviet Union, which he saw as Anthill, he named it Admass. Urban folklore tells us that all airports, and most cities, with their faceless modernist and post-modernist towerblocks, look alike. In terms of place authenticity, British Columbia now sports a Bavarian-style town and a Spanish-style town, and more settlements are being done over in equally inauthentic styles for the tourist market. I'm moved to laughter by the replica of Ann Hathaway's Cottage in Victoria, British Columbia, which once bore the damning recommendation: '"most authentic" – Walt Disney'.

The resulting landscapes may fairly be compared with Muzak. In Muzak the original music, whether Bach or the Beatles, 'is rearranged, rerecorded and electronically processed so that all the major variations in tonal range, in noise level and in rhythm are compressed into a narrow band' (Relph 1982). Thus melodies somehow all sound the same because all the extremes and idiosyncrasies have been deleted. In a similar fashion, fast food is processed and homogenized to produce the identical taste, acceptable to all, across the length and breadth of North America. When landscapes become so homogenized, Relph (1982) calls them 'flatscapes', while I regard them as 'bland-

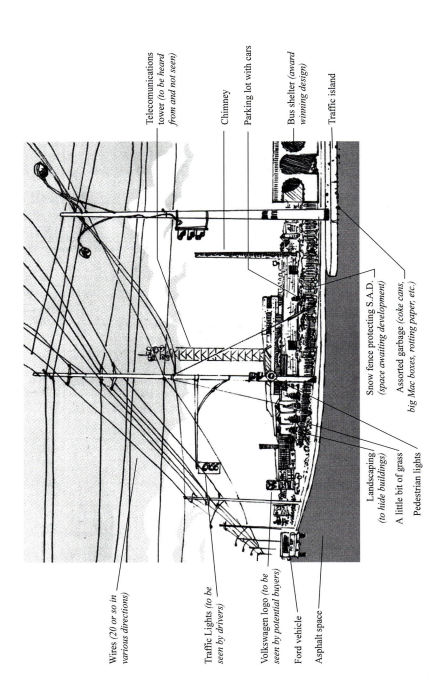

Wires *(20 or so in various directions)*

Traffic Lights *(to be seen by drivers)*

Volkswagen logo *(to be seen by potential buyers)*

Ford vehicle

Asphalt space

Telecomunications tower *(to be heard from and not seen)*

Chimney

Parking lot with cars

Bus shelter *(award winning design)*

Traffic island

Snow fence protecting S.A.D. *(space awaiting development)*

Assorted garbage *(coke cans, big Mac boxes, rotting paper, etc.)*

Landscaping *(to hide buildings)*

A little bit of grass

Pedestrian lights

Figure 4.1 A modern nowhereland (after Relph 1982)

scapes'. (Figures 4.1 to 4.4). This homogenization of landscape, an efficient rush toward deadly dullness, is particularly alarming when we realise that diversity is necessary for mental well-being (Chapter 3).

It is Relph's opinion that the source of this 'ugliness in modern landscape lies deep within the personality of our technological and consumer society' which, in the decision to found a society solely on economic grounds, has sunk into a state of 'visual illiteracy'. We are, in fact, living in the state prophesied by T.S. Eliot (1948: 19): 'a period of some duration of which it is possible to say that it will have no culture'. Lewis *et al.*'s (1973) discussion of visual blight in the United States provides an example of the results, in terms of landscapes, of a culture without culture.

But these rather gentle critics, mainly geographers, are loudly overborne by a far more raucous chorus of aesthetic disapproval on the part of architects and urban designers. Planning critique (Porteous 1989) has a long history, and one of its major components is the reasoned critical analysis, often expressed in terms of the grossest outrage, of modern urban and rural environments.

Although designers have directed diatribes against the townscapes of Britain, France and Australia, the majority of the critical invective has been inspired by the cities of the United States. The Magna Carta of the anti-ugly campaign was the December 1950 issue of the London *Architectural Review* which inveighed against 'The mess that is America.' The charge was then taken up by Americans themselves.

The 1960s was the 'decade in which Americans first turned a deliberately

Figure 4.2 Blandscape in Littlehampton, Sussex: uniformity, no trees or chimneys, pretentious door-cases and windows

Figure 4.3 Anywhere UK, 1960s. Actually Hull, Yorkshire 1965

aesthetic eye on the land they live in' (Lowenthal 1979). Alsop (1962) in 'America the ugly' spoke of 'a subconscious consensus that there is something manly in messiness and ugliness, something sissified about whatever is handsome, or well-ordered, or beautiful' (Figures 4.5, 4.6). Udall

Figure 4.4 Anywhere in the world, 1990s. Actually Havana, Cuba, 1995

Figure 4.5 Canadian version of 'America the ugly', Alberta

(1963) included visual blight in his analysis of *The Quiet Crisis* of environmental degradation, while Tunnard and Pushkarev (1963) deplored the glacierlike spread of amorphous suburban blandscapes and demanded 'an environment worthy of man'.

Perhaps the most compelling assault, however, was the tellingly-entitled *God's Own Junkyard* (Blake 1964), written not in anger 'but in fury'. Blake illustrates 'the planned deterioration of America's landscape' by junkyards, wirescapes, billboards and ugly commercial signage, and shows how this ugliness is actually 'planned' by outdoor advertisers, real-estate brokers, automobile manufacturers, subdividers, and utility corporations. He asserts that there are no attractive cities in the United States and that 'in America today no citizen (except for an occasional hermit) has a chance to see anything but hideousness all around him, day in and day out'.

Outside observers came to much the same conclusions. The prolific and acerbic landscape critic Ian Nairn (1965), known as the Don Quixote of environmental journalism, found the American landscape 'full of goop ... a chaos of nonrelation'. Being used to the more 'natural' curvilinear landscapes of Europe, he found the Midwestern scene especially awful: 'a set of ruled squares ... filled ... with low-intensity muck'. And although the major thrust of the anti-ugly campaign came in the 1960s (Lynch 1960, Jacobs 1961, Mumford 1964), American diatribes continue (Kunstler 1993) and among foreign observers, the perceptive Hungarian poet George Faludy (1988) was moved to remark that 'only in America is mediocrity ... raised to the level of cult and positively admired' (p. 44). His reasons for the banality of American culture and landscape are worth listing:

Figure 4.6 Canadian version of 'God's own junkyard', British Columbia

Such *Kultur* as our age possesses is pre-eminently contemporary Anglo-Saxon, i.e. it is technocratic, money-minded, and power oriented (p. 92) [resulting in] the mindless auto-eroticism of people who have been deprived by urbanization, mass media, and spiritual uprootedness of their entire cultural inheritance .... The proletarian is the man or woman or child who has been robbed by fate of a rich inheritance and has learned to be content with the dregs of a debased, mass-produced culture, whether he lives in a slum or (as frequently happens) in the White House (p. 22).

We are all proletarians now.

More analytical in his concern with landscape, Lowenthal (1985) suggests that the common themes which underlie landscape ugliness are: defects of content (scenes spoiled by ugly objects, for example); defects in order and arrangement (beautiful objects not harmoniously arranged); defects of vision (landscapes generated by selfishness, greed, and ignorance); and defects of social value (where environmental inadequacies reflect, as Faludy points out, deep social malaise).

The social malaise, however, is seen by some as less aesthetic than ethical. For both Jane Jacobs (1961) and J.B. Jackson (1964) assert that 'a city is not a work of art' but a place for living and working, and that liveability should take precedence over aesthetics. Gardiner (1971) speaks of 'cruel aesthetics', whereby architectural elites impose their designs on the public without consultation, with the result that citizens find themselves in grossly user-unfriendly buildings, some with no windows, for example, and others with so many that occupants overheat, pull the curtains and thus lose their panoramic views. This anti-aesthetic backlash, however, was really a polemic in favour of citizen action and participatory planning, as has been detailed in Porteous (1977). There need be no difficulty in reconciling liveability with aesthetics; indeed, the latter is clearly a subset of the former.

The aesthetics of townscape returned to the fore in the 1980s, largely because of the efforts of Ada Louise Huxtable of New York and H.R.H. Charles, Prince of Wales. Huxtable, architectural critic for the *New York Times*, composes biting essays on the metropolitan landscape, some of which were reprinted in the aptly-titled *Will They Ever Finish Bruckner Boulevard?* (1970). A more recent polemic, sporting the engaging title *Goodbye History, Hello Hamburger* (1986) demonstrates the continuing loss of architectural heritage in the United States, and its replacement by the built equivalent of the homogenized, basically tasteless, hamburger patty. Huxtable is amazed that Americans rejoice in the visual sameness that stretches from New York to Los Angeles. This joy in the bland and the mediocre certainly points to national immaturity, conformism, and fear of the really, rather than super-ficially, different.

Prince Charles is somewhat more staid than Huxtable. Amazingly activist for a member of the British royal family, he stirred a hornet's nest in 1988 with a prime-time television programme on the banality and sheer ugliness of recent large-scale architectural development in Britain, and particularly in London. The programme was followed by a public exhibition illustrating the Prince's views of good and bad townscape, and finally by a book, *A Vision of Britain* (1989).

Charles is rather conservative, and prefers a classic style of architecture. Nevertheless, he convincingly demonstrates how the arrogance of avant-garde architects, coupled with the sheer greed of real estate developers and big business, has served to reduce enormously the aesthetic pleasure obtained from the London townscape by resident and visitor alike. Both London and Paris have been utterly changed in the last thirty years by a series of unsympathetic developments. In particular, whole skylines have been ruined and major aesthetic symbols, such as St. Paul's Cathedral, obscured by undistinguished, out-of-scale towers which would look much better in Kansas City or Singapore (Figure 4.7). The traveller seeking an idea of what London and Paris were like before greed took hold must now journey to the relatively unspoilt central townscapes of urban backwaters such as Prague and Rome.

It is clear from the foregoing that many architects and most corporations and developers are sensitive not to place but to ego and pocketbook. They see no reason why every major city should not be made to look like New York or Los Angeles. From both liveability, or 'live-in', and aesthetic, or

Figure 4.7 London's spires and domes increasingly obscured by identikit towers and blocks

'look-at', points of view, it is high time that citizens took steps to prevent the desecration of their heritage and found ways to monitor and control all sizeable new developments.

4.2 CITIZEN ACTION

Citizen action can undermine authority, erode power, bring down governments, and stop a coup dead in its tracks. Action in the aesthetic realm, however, suffers from the paralysing problem of elitism.

The problem of elitism

The chief issue in activist aesthetic politics is that public aesthetics appears to be taken seriously only by a minority. This minority assumes that its taste is preferable, and seeks to impose it on both the apathetic masses and, more importantly, on the unaesthetic but powerful minorities who run big business, government, and planning agencies.

The problem was brought to the fore in an outspoken article by Wildavsky (1967), provocatively entitled 'Aesthetic power or the triumph of the sensitive minority over the vulgar mass'. Wildavsky noted that the conservation movement began with a concern over the feared disappearance of vital resources, moved on to ecology, pollution, and recreation values, and only thereafter began to consider aesthetics.

Once aesthetics is admitted to be important, we have the cheerless spectacle of resource economists trying to price intangibles, such as views. Public preference surveys are not necessarily useful, because they are generated in a passive laboratory or questionnaire situation; there is no guarantee that these attitudes are based in action or will ever result in action. Further, there seems no way of ensuring that the revealed preferences are in any way meaningful, unless they can be shown, as in the Ulrich work, to have physiological correlates. Finally, no agreed-upon set of rules exists for translating preferences into public policy. And Wildavsky's 'vulgar mass', although its consciousness has been raised enormously since the anti-ugly campaigns of the 1960s, still values other gratifications more highly than the aesthetic quality of environment. Surveys which ask the American public what improvements they would like in government programmes rarely reveal any interest in aesthetic issues (Wildavsky 1967).

It is clear, then, that given prevailing public indifference, the aesthete minority which vociferously argues for improved townscapes and more wilderness preservation is asking government to subsidize its minority predilections. As so many studies have shown, this minority is likely to be white, highly-educated, affluent and middle-class. A clear case of elitism, one would think. Besides, before we act on aesthetic issues, surely more knowledge is needed about this obscure area (the same 'more research is needed' argument used by Reagan to slow down American government response to acid rain in the mid-1980s).

The counter-arguments are as follows. Conservatives do not concede that democracy raises the average cultural level of a community; they stress instead its debasing effect. Modern democracy appears to be antithetical to aesthetics, for it is 'partial to the average and hostile to the exceptional ... Ben

Franklin once described America as a land of "happy mediocrity"' (Tuan 1993: 209). Thus it 'remains perfectly acceptable to lament the degradation of public or aesthetic discourse' (Maier 1992: 130), with the result that aesthetics, like the money supply or school curricula, continues to be an area of collective life to be stoutly shielded from majority rule.

Liberal arguments are a little more generous. Elites have always existed; in most cases it is they who have been responsible for changing the world, for better or worse. The 'trickle-down' theory of culture demonstrates that what is elite, minority taste in one era often becomes general, majority taste in a later one. Thus the North American, British, and Australian predilection for detached houses with front lawns and enclosed backyards mimics, in little, the eighteenth-century mansions and parks of the rich. Further, given the rapidity of change in a world of future shock where all that is solid melts into air (Berman 1982), there is no time to wait for aesthetic environmental education to reach the masses. Secure in the knowledge that what they prefer is likely to be preferred later by the culture in general, aesthetic activists cry: 'Act Now!'

And to the obstructive argument that they are acting on inadequate knowledge, the activist answer is that all knowledge is provisional. They point to the plight of Dostoevski's 'underground man' who so appreciated the endless depths of every side of every question that he became totally unfit for any form of action. Activists, then, are willing to shelve questions of inadequate knowledge and elitism in the face of the increasing threats to environment from rampant change in a technologically-powerful but often aesthetically and ethically bankrupt culture. As Wildavsky (1967) forcefully states, in relation to the 1960s push for wilderness preservation: 'Weak and frail as we are, beset by doubts and anxieties, undoubtedly partial in our views, we must act. If those who love the wilderness will not save it, who will?'

Reactions to the debate were mixed. Haskel (1965) thought that President Johnson's 'national drive for beauty' might well result in 'sentiment and snobbery', especially if professionals' hatred of 'popular taste' becomes enshrined in legislation. He pointed out that the American public likes automobiles, suburbs, and parking lots. Swamps and wetlands were once disdained, but are now seen as valuable ecosystems. Similarly, might not junkyards and billboards become regarded as valuable assets? Lewis (1970) agreed that the American landscape is conditional upon deep-seated popular tastes for convenience and information. Further, he argued that American attitudes in favour of 'freedom', however unfounded, result in beliefs that beauty is too costly and that governments should not dictate taste. 'Who's going to tell me what to like?' is the cry, which thoughtlessly forgets that such is the chief function of advertising. Lewis's conclusion was that we must either learn to enjoy ugliness, learn to ignore it, or cast caution to the winds and try to change it.

Political conservatives (who, ironically, often bely their label in their desire for 'progress' and change) were represented in the debate by William

F. Buckley (1966), who predictably cast aesthetic and ecological activists into the moulds of either 'the very gloomy' or 'the *el fastidiosos*' who inevitably prefer European ways to American. The latter always touches a raw nerve in a nation stupid enough to demand that one should 'love it or leave it'. More constructively, Buckley agreed that tastes differ, but noted that not all tastes are equally defensible. He suggested that the function of government be primarily negative, in that certain grossly unpleasant practices should be prohibited and others, such as billboards, be regulated because of the public's right to protection. He reluctantly concluded, in the face of general indifference, that 'the only way to do anything ... is to do so athwart the people's indifference; indeed, by extension, against their will'.

The era known as 'the Sixties' was valuable in changing public attitudes on a wide range of topics. It is clear that the anti-ugly debate of that era served to educate the general public to the aesthetic aspects of environmental problems. By the late 1970s it was possible to say that although aesthetic taste was still defined by elites, 'their perception ..., with time, has become our perception' (Lowe 1977). In a world of affluence and mass communication, trickle-down time, of course, is very fast. The success of the campaign in North America can be measured in the wealth of heritage preservation in cities, wilderness conservation, the re-creation of prairie grasslands and a host of similar movements. In Britain, the rash of guides to the countryside, the popularity of books on landscape history, and the sudden revival of long-distance walking in the 1980s evinces the success of the movement. In fact, so popular have country walking in Britain and wilderness hiking in North America become, public demand for these aesthetic experiences has become a major threat to the integrity of the landscapes in question. Part of the justification for the recent boom in recreation research and park management is the need to protect sensitive environments not merely from rapacious corporations but also from a new generation of aesthetes!

Direct action

On this basis citizen action in aesthetic matters, unusual in the 1960s, had become fairly common by the 1980s with the rapid growth of aesthetic advocacy groups and of tactical guides to lobbying (Ridout 1988). Citizen action, by definition, is sometimes illegal and usually extra-legal, involving trespass, restriction of the movements of others, and disruption of work schedules.

One reason for direct action is the general failure of the 'public participation in planning' movement. Despite heroic efforts on the part of planning activists and critics, public participation in planning generally remains on the lower rungs of Arnstein's (1969) ladder of participation. Too often the public is appeased, co-opted, or simply ignored.

Direct action, in contrast, forcefully brings a group's concerns to the attention of both the public and the relevant authorities. It can be dangerous but is often exhilarating. Its chief successes have been in halting massive developments, notably the building of unnecessary freeways, airports, and supersonic passenger aircraft. As early as 1971 there were over 350 disputes involving highway building in the United States alone.

Tactics are almost invariably passive. People chain themselves to historic trees or buildings, stand or lie in the way of bulldozers, wave placards, demonstrate outside government offices, harass despoilers, and lobby politicians (Figure 4.8). Such tactics, in combination with reasoned argument in public meetings, letter-to-the-editor campaigns, and the formation of interest groups which capture media attention, have had quite good results across the Western world, including wilderness preservation, park creation, farmland landscape subsidies, neighbourhood downzonings in favour of single-family houses over high-rises, the undergrounding of wirescape, subsidized paint-up schemes, conversion of older buildings instead of demolition, 'heritage building' designations, strict sign by-laws, the preservation of street views, and an exponential growth of street art on what were formerly dull blank walls (Figure 4.9).

Studies of activist profiles reveal (Singer 1979) that by the late 1970s activists were no longer simply a white-collar elite. The Waterfront Coalition of Hudson and Bergen Counties, for example, was formed in 1977 after several Not-In-My-Backyard (NIMBY)-style battles to save the New Jersey waterfront opposite Manhattan. Many of the activists were blue-collar, lower-income, and sometimes unemployed. Although most of the leaders showed the typical middle-class activist profile, they were able to

Figure 4.8 Environmental protest in Philadelphia, 1970s (courtesy Roman Cybriwsky)

Figure 4.9 Wall art, Victoria, BC

mobilize whole neighbourhoods with no difficulty, resulting in public meetings of up to one thousand people opposed to the erection of 'noxious facilities' in 'unspoiled areas' in likely contravention of the Coastal Zone Management (CZM) Act of 1972. The coalition proved hugely successful because of the large numbers of protesters, the broad spectrum of social classes involved, and clever use of the mass media. One of their proposals was not NIMBY-related; they sought to revitalize the CZM Act to place greater emphasis on 'ecological, cultural, historic, and aesthetic values'.

Ecotage

Citizen action reaches its extreme in ecotage, which is defined as sabotage in favour of the environment. The reader interested in the theory and practice of ecotage is directed to the many books of the anarchist philosopher Edward Abbey (Hepworth and McNamee 1985), and in particular to essay collections such as *The Journey Home* (Abbey 1977) and *Abbey's Road* (1979) and the novel *The Monkey Wrench Gang* (1985). Briefly, the concept involves the selective disempowering of the tools used in environmental destruction. In *The Monkey Wrench Gang*, which appears to be a sabotage manual as much as a novel, the characters discuss the origin of planned, legal, ecological disasters and go about sabotaging the machinery of the industrial corporations held responsible. As everyone knows by now, a little sand or sugar, judiciously introduced, can bring the largest yellow earth-mover to a grinding halt. Spikes hammered into trees naturally discourage the attentions of loggers, who are informed beforehand of the danger.

Ecodefense: A Field Guide to Monkeywrenching (Foreman 1985) makes interesting reading for its detailed sections on tree-spiking, flattening tyres, burning machinery, 'vehicle modifications', fence-cutting, trash return, disrupting illegal activities, jamming locks, billboard-trashing, billboard-burning, billboard revision, spray painting, camouflage, self-defence, pursuit and evasion, and finally, arrest. Edward Abbey's 'Forward!' to the book ably and, I believe, correctly, trashes 'the three-piece-suited gangsters [of] the international timber, mining and beef industries [who are] invading our public lands ... bashing their way into our forests, mountains, and rangelands and looting them for everything they can get away with, ... actively encouraged by those jellyfish government agencies which are supposed to protect the public lands, and as always aided and abetted ... by the quisling politicians of our Western states', whom he goes on to name. Abbey is intemperate (Petersen 1994), but he is right, for right-wing capitalism is not compatible with sustainability. It is because the greedy and powerful rule the earth, and are visibly destroying it, that ecotage becomes necessary.

Ecotage can clearly be both expensive and dangerous, but so can unthinking ecological destruction in the pursuit of profit. Ecotage proponents deny that their work is ecoterrorism, for they expressly avoid violence against people and other living things (such violence is best left to governments and corporations). On the contrary, sabotage is merely 'the application of force against inanimate property'. Indeed, while Greenpeace, for example, has been guilty of harassing French nuclear testers in the South Pacific, it fell to the French government to draw first blood by killing a Greenpeace volunteer in New Zealand in 1985.

Further, sabotage is regarded as a last resort, to be deployed only when all else fails, and is seen as legitimate defence in favour of both capitalism's and communism's progressive destruction of nature. The Earth First! organization, often accused of being responsible for ecotage actions, explicitly disavows them. Earth First! does not advocate such actions, for 'monkey-wrenching is a personal matter'. Further, the organization denies any philosophical underpinnings with the slogan 'talk is cheap, action is dear'. Such denials are absolutely necessary if ecoactivist organizations are not to be held collectively responsible for ecoactions involving the loss of corporate property.

Few activists engage in ecotage. Yet the range of activities open to the environmentalist aesthete is extremely wide and mostly legal. One of the more promising was the growth of Green parties in Europe during the 1980s. Such parties have been successful in attracting up to 10 per cent of the vote in German and Dutch elections and, although unlikely ever to gain power, they have had the satisfaction of seeing some of their policies co-opted and legislated by more traditional political parties.

4.3 LEGAL ISSUES

The obvious short-term desire of the ecoactivist or citizens' action group is to prevent a particular development or to blunt its worst excesses. Their longer-term strategy involves the enshrinement of general aesthetic and ecological principles in legislation at local, regional, and national levels.

It is instructive at this point to trace the gradual inclusion of the concept of beauty in American legislation. It was only with the general adoption of planning and zoning practices in American cities in the early twentieth century that the law began to take cognizance of beauty.

Chandler (1922) asserted that the chief problems in the legal consideration of beauty were the general beliefs that 'beauty is in the eye of the beholder', that 'there's no accounting for taste', that 'beauty is only skin deep', and the like. That these beliefs are largely erroneous has been demonstrated in two earlier chapters. Nevertheless, the American ethos in the era of raw capitalism known as 'the Twenties' was one of unabashed individualism, and law, like planning, cannot be individualistic: 'Law is the sum of rules imposed upon the individual to which he must conform whether he likes it or not. Law is the limit beyond which the self-expression of the individual is not permitted.'

Nevertheless, aesthetic considerations have been enshrined in law since Roman times, and especially in relation to the uncontrollable senses of sound and smell. Late nineteenth-century zoning practices developed in response to problems of urban health, safety, and the senses of smell (e.g. glue factories) and sound (e.g. shipyard riveters). Yet by the 1920s no such consideration had been given to the sense of sight, for United States courts repeatedly upheld the dictum that 'the police power cannot be exercised for aesthetic purposes' (Chandler 1922). This ruling was used, for example, to strike down state laws forbidding the erection of billboards within 500 feet of parks and boulevards.

Beauty sneaked into legislation by the back door. The development of urban parks at the turn of the nineteenth century was expressly meant to serve 'not only the grosser senses, but also the love of the beautiful in nature'. In some jurisdictions, like Chicago, the prohibition of billboards, originally developed for aesthetic reasons, was upheld by the courts on the legally-sound grounds of public safety. In residential zoning, the enhancement of the aesthetic content of subdivisions was a by-product of zoning rules designed to prevent 'inharmonious intrusions'.

The major problem in this pre-social science era was a lack of standards: 'there was no standard by which objects offensive to the eye could be judged'. A related problem was that of trying to create legislation which reflected public taste and yet did not interfere with the unfettered taste of the individual. Chandler's optimistic conclusion was that, despite these fundamental drawbacks of the culture, future decades would see beauty 'recognized, protected, and fostered by law' (p. 474).

Fifty years later Broughton's (1972) review of the methods by which aesthetic values might be protected demonstrates that the major thrust for aesthetic legislation came only with the ferment of the 1960s. The vast increase in environmental protest in that era clearly signalled a fundamental unease about the way 'progress' was being measured. Significant sections of the population began to question The Pig Principle ('more is better') and to consider a lifestyle variant of the Bauhaus architectural dictum that 'less is more'.

Consequently, the rapid increase in court actions and legislation from the 1960s onwards brought aesthetic issues to the fore as a component of environmentalism. Indeed, Broughton claimed that aesthetics could be regarded 'as a proxy or even ... an indicator of environmental quality in general', while many ecological movements, such as those to save the bald eagle or preserve wilderness landscapes, were seen as 'proxies for the effort to preserve the aesthetic'. By the late 1960s much effort was being made to reach some conclusion regarding the nature of the citizen's assumed right to aesthetic quality in the public environment.

During the 1970s several landmark cases demonstrated an increasing willingness on the part of Supreme Court judges to support the rights of local governments to exert police power to preserve aesthetic wealth. The classic cases include *Penn Central Transportation Company* v. *City of New York*, 438 US 104 (1978), *Metromedia, Inc.* v. *City of San Diego*, 453 US 490 (1981), and the Los Angeles case known as *Member of City Council* v. *Vincent*, 104 S. Ct. 2118 (1984). Respectively, these aesthetic legal cases involved an attempt to build an office tower above historic landmark-designated Grand Central Station, the regulation of signage, and the specific prohibition of the posting of signs on public property (Pearlman 1988).

What is important here is that the Supreme Court has now become strongly supportive of local government efforts to control the aesthetic quality of environment. The Court feels that zoning for aesthetics serves a valid public purpose, providing not only economic benefits but also adding to the quality of urban life. Further, an analysis of similar cases at the state level confirms that 'aesthetic considerations are seen as primary and ... non-aesthetic aspects as derivative' (Pearlman 1988: 489). Finally, the courts do not appear to require that the aesthetic judgements of city governments be justified on 'objective', research-based grounds. It is sufficient for a local government to express its preferences and attitudes in the form of aesthetic legislation. Indeed, at the state level, at least eighteen jurisdictions accept environmental regulation based on aesthetics alone and only nine states specifically prohibit regulation based solely on aesthetics.

In sum, the general trend in the 1980s was toward an increased acceptance of regulations based on aesthetics. The result of this is legislation, which in aesthetic terms may usefully be categorized as nuisance law, zoning law, and administrative law.

Nuisance law

Private nuisance law is controlled by the judicial application of principles of common law. A private nuisance is defined as 'an unreasonable interference with another's enjoyment of real property'. The term 'unreasonable' depends both on the values of judges and on their assessment of what might be unreasonable from the point of view of the average citizen. Most cases involve not merely claims for damages, but the active seeking of injunctions to stop the nuisance.

Thus where ugliness is the issue the court must be convinced, first, that the decrease in aesthetic value is substantial enough for the court to take cognizance, and second, that the benefits of an injunction will outweigh the detrimental effects to the community. In practice, aesthetic nuisance cases rarely aim to preserve beauty but rather to prevent unusual ugliness.

Cases are frequently lost because of the power of economic interests, the perception of aesthetics as a luxury, and lack of agreed-upon standards. When a case is won, it is often won on non-aesthetic grounds, although offensive sounds and smells often result in the successful granting of injunctions. Legal opinion, however, has increasingly begun to consider that there is no reasonable basis to divide sight from the other senses, and that standards of visual ugliness which might well offend the average citizen should not be too difficult to ascertain. In this way visual aesthetics shares the definitional problems of obscenity law.

Zoning Law

There are far too many inherent problems in nuisance law for it to be a useful method for the preservation and enhancement of beauty. Zoning laws, which directly control public conduct, stand in an intermediate position between nuisance law and state or national government legislation which requires action. Zoning by-laws are enacted by urban or regional authorities.

Most early aesthetic zoning by-laws involved attempts to regulate the rash of billboards which in the early twentieth century threatened to make the American landscape resemble a scene from the film *Brazil*. The general tenor of these early by-laws was that police power was appropriate for health, safety, or moral purposes, but not for aesthetic purposes. Thus those who opposed billboards on aesthetic grounds were forced to argue against them on grounds of health (they shut out light), safety (they serve to collect inflammable rubbish around them) and moral standards (they serve as useful screens for those who wish to commit public nuisances).

Only in the 1950s did American zoning by-laws begin to acknowledge that the aesthetic criterion (billboards are ugly!) might be acceptable to the general public. By 1964 a New Jersey court felt able to opine that 'the aesthetic impact of billboards is an economic fact which may bear heavily upon the enjoyment and value of property', and soon thereafter, which not coincidentally was the era of the vociferous anti-ugly campaign detailed above in Section 4.1, similar principles were being applied to junkyards, architecture, landscaping and open space. 'God's own junkyard' was at last being dealt with.

Yet an analysis of architectural aesthetics-control mechanisms in the United States, Canada, Britain, Australia and New Zealand (Preiser and Rohane 1988) revealed that the Junkyard was being dealt with in very different ways in different jurisdictions. Controls ranged from non-existent in Houston and Amarillo, Texas, to a very large number of regulated items in Britain. In general, Commonwealth countries tended to be influenced by the long tradition of aesthetic controls in the United Kingdom. In the United States, older, eastern cities tended to be more regulated than newer, western ones.

Zoning law, therefore, is likely to be a useful indicator of the aesthetic standards of the community. For nationwide or regional common standards on items of general interest, however, it is clear that legislation by bodies superior to the metropolitan level is required.

Administrative law

Laws enacted by regional (state, province or county) or national legislatures and applicable throughout a major jurisdiction are the preferred legal means of compelling positive aesthetic action. They are relatively much more easy to impose than nuisance or zoning law, and can be enforced quite readily, especially in relation to the large-scale developments frequently envisaged by public or semi-public agencies such as the US Corps of Engineers or the US Forest Service.

When environmental impact statements are prepared, however, it remains easy to ignore or downplay aesthetic issues. When generating cost-benefit analyses, it is not difficult to undervalue the important considerations which economists have long called 'intangibles'. Cost-benefit analysis, of course, presupposes the existence of a market, which is not always relevant in aesthetic terms, and can also be readily manipulated and fudged. It is not uncommon for dam-builders to assert that the resulting lake will far surpass in beauty the landscape that they propose to drown.

Because of the importance of aesthetic legislation at the national and regional level, it will be discussed in some detail in Section 4.4 below. This discussion of legal issues in aesthetics concludes with case-studies of classic legal battles fought over aesthetics in both urban and rural areas of the United States.

Urban aesthetic warfare

Brace (1980) provides a useful review of urban aesthetic cases of the 1970s, beginning provocatively with the statement that 'the legal profession ... is known more for its readiness to protect the rights of individuals to profits than the rights of the public to a pleasing environment'. As early as 1954 the

Supreme Court had opined that 'it is within the power of the legislature to determine that the community should be beautiful as well as healthy', yet vested interests apparently succeeded in quashing effective aesthetic action in cities during the following decades.

From the mid-1970s, however, several major suits were launched on aesthetic grounds alone. Typical examples included moves to prevent the development of high-rise towers which would inevitably spoil existing city skylines. The most spectacular case was the decision made by the Supreme Court on 26 June 1978 which decided for the City of New York against the Penn Central Company. Under its Landmark Preservation Law, the city's move to prevent Penn Central from building an office tower on top of Grand Central Station was upheld. The issue was not merely one of historic-landmark preservation but also concerned the aesthetic effect of a fifty-three-storey addition to the low-rise station. In effect, this Supreme Court decision affirmed the rights of cities and states to protect their aesthetic character.

In the same year the State of Washington Supreme Court upheld the City of Seattle's refusal to grant permission for a condominium development. The grounds for refusal were chiefly aesthetic, for the building conformed to existing zoning regulations and would not have led to unacceptable increases in noise and traffic volume, nor caused major reductions in adjacent property values. Rather, the proposed building was held to be of excessive bulk, out of scale with regard to its surroundings, and likely to block sunlight, create major shadows, and obstruct views.

From the six cases reviewed, Brace (1980) concluded that aesthetics, by 1980, could at last be considered a legitimate concern of governments because of its recognition as an important component of the urban quality of life. Further, although standards remain a matter of taste, public taste may readily be ascertained via hearings, meetings, surveys and the like. Finally, although aesthetic values will never be precisely quantified, they can readily be stated as public policy goals and applied reasonably well.

The LILCO case

In the mid-1970s a series of hearings was held by the New York State Public Service Commission to determine if the Long Island Lighting Company (LILCO) should be required to put underground a proposed transmission line through the town of Riverhead (Carruth 1977, Gussow 1977). This is the classic rural aesthetics legal case, and is worth considering in some detail for the arguments employed by the two sides.

LILCO argued for an overhead transmission line, stating that the cost of going underground was unwarranted given the local landscape of flat potato-growing farmland with no unique scenic qualities. Moreover, the community was so small that very few people would ever see the transmission towers or poles.

Local citizens of Riverhead and Suffolk County countered by suggesting that the impact of the transmission line would be severe. To support their argument they retained the services of an expert witness, the painter Alan Gussow, described as an 'expert of landscape perception'. Gussow gave in total 150 pages of testimony.

He described the visual character of the area as serene, primarily natural, and visually both harmonious and restful. Its two major features are a strong horizontal plane, in which existing 25 and 69 kV power poles do not obtrude above the trees, and the existence of vistas confined by these same trees.

He argued that the intrusion into this landscape of about 150 345 kV towers each 150 feet high would have a galvanizing effect 'equivalent to a telephone call in the night'. The towers would overwhelm all other landscape elements, including the tallest trees, their insistent verticality would severely impinge on the existing dominance of the horizontal plane, and their intrusion would spoil many confined vistas which are currently visually attractive. In short, such rampant verticality is out of context in flat farmland and would destroy the area's visual harmony and human scale.

LILCO countered with its own expert, David Carruth, a landscape architect and consultant to the company. Carruth argued that although appraisals of scenic quality are complex and difficult, most existing methods (which will be reviewed below in Chapter 5.1) agree that landscape quality increases with relief, topographic elements, and visual complexity (see Chapter 3.2 above). In other words, 'the greater the topographic variation, the greater the scenic quality', and Riverhead, being flat, was therefore necessarily of low quality in aesthetic terms. Ground level variation was shown to be less than 100 feet (the 'lowest category' of scenic quality). There were few viewpoints, and a severe lack of the diversifying elements that yield complexity. Further, there were no 'unusual' or 'unknown' elements to prevent boredom.

Gussow at that time was not aware of the culture-bound and intellectually specious character of contemporary landscape assessment methods (Chapter 5.1). Nor did the Riverhead residents emphasize that Carruth's arguments pertained wholly to the visitor's point of view, nor that they had an attachment to and love of the land as it then was (see Figure 1.3). Indeed, LILCO argued that there was no known way to incorporate artistic opinion or local feeling into a 'scenic quality rating for public consideration'. This dismissal of local sentiment is an extremely common phenomenon, the usual argument being that national or regional interests must necessarily override local ones (Porteous 1989). Instead, LILCO completed a landscape assessment for the whole proposed line, providing numerical data on the 'visual impact' measured from 'the travelling public's point of view'.

The wily Gussow refused to be caught out on a series of sticky points. He refused to discuss the cost issue on the grounds of lack of qualification, asserted the artist's experience and intuition to be as valid as supposedly objective landscape measures, and clarified the point that his 'subjective' opinions were not merely capricious, but based on years of landscape assessment experience.

His most telling point lay in the realm of symbolic aesthetics. The farm landscape, he averred, has integrity, and is a valuable natural resource. Ordinary landscapes are as important to our sense of beauty as spectacular scenic wonders (Meinig 1979). It is useful for city children to be able to experience unspoiled farmland so close to New York City. Most important, however, is the symbolic value of farmland in the United States. Because of the historic development of the nation from a farming foothold in the Northeast, farm landscapes have important historic, cultural, and mythic content and, indeed, may be said to be the American equivalent of a Greek temple or Chartres cathedral.

The latter proved to be a telling blow, but the importance of the case does not wholly lie in its outcome. The LILCO case is classic in that it clearly points up the issues already developed in the Humanists and Activists chapters and which are discussed further within the Planners chapter below. Specifically, these issues concern the general disdain of planners and developers for local residents, the lack of importance granted to attachment or love of landscape on the part of residents, the importance of expert testimony, and the classic struggle between the intuitive expert and the numerate methodologist, between artist and social scientist.

4.4 PUBLIC POLICY

It is interesting to note that both legal cases and national legislation have a tendency to consider issues of fittingness, congruity, and harmony when dealing with landscape aesthetics. These components of aesthetic value are, of course, the same attributes discussed by humanists since Aristotle and confirmed in their importance by much experimental work. The challenge comes when these aesthetic concepts must be translated into workable public policy at the national level. The problem is discussed below in case-studies of the United States and Britain.

The United States

Zube (1976a, 1980) has traced the development of the process whereby aesthetic issues have become an important aspect of environmental public policy. The earliest movement, typically American in its concern for the vast, huge, and unique (see Chapter 2.4 above) was the late nineteenth-century concern to preserve unique landscapes. One of the very first of these 'scenic wonders' to be preserved was the Yosemite Valley in California. As early as 1864 the US Congress declared that 'it is the will of the nation that this scenery shall never be private property but ... shall be held for public purposes'. Yosemite was followed closely by Yellowstone National Park (1872) and by 1916 Congress had enacted legislation creating fourteen national parks each valued for its uniqueness.

The basic mandate of the National Parks Service, also created in 1916, was to preserve scenery, with public use as a secondary issue. But the enormous growth of outdoor recreation from the early 1950s forced a shift in management attitude embodied in the work of the Outdoor Recreation Resources Review Committee (ORRRC). Having discovered that driving and walking for pleasure were the two most frequently reported recreations, the ORRRC promoted programmes creating national scenic roads, scenic trails, and the National Wild and Scenic Rivers Act of 1968, which designated rivers according to their scenic, recreational, or environmental values. The chief value here is less preservation than the enhancement of the recreation experience. One spin-off from this movement was a major research effort on the relationship between perceived landscape quality and the physical attributes of landscapes, the theoretical aspects of which have been reported above (Chapter 3) while the applied aspects are discussed below (Chapter 5).

Only with the 1960s came stage three, the identification of 'scenic misfits' and the amelioration of the ugly. Stimulated in part by the invective of the literary activists (section 4.1 above), the 1965 White House Conference on Natural Beauty seriously addressed the loss of landscape beauty and was as much concerned with the elimination of the ugly and the restoration of beauty as with the protection of natural environments. This movement went hand-in-hand with the heightened aesthetic awareness of the public caused by the vociferous involvement of activists in both protest and legal cases.

Achievement in this area has never matched the success of stages one and two, chiefly because of the opposition of entrenched economic interests, particularly in forestry, mining and outdoor advertising. Nevertheless, significant changes have occurred, such as Vermont's highway billboard legislation, the incorporation of 'visual resource management' training for Highway Administration engineers, legislation in 1977 to ensure the restoration of landscapes ruined by strip-mining, and the involvement of landscape architects in the development of aesthetically acceptable logging practices by the Forest Service.

By the late 1960s, however, it had been realized that stages one to three were piecemeal approaches, given that the preservation of unique 'wonders' and the prevention of scenic misfits are merely the extremes of an aesthetic continuum, and that all the more ordinary landscapes in between these extremes deserved consideration. The result was the National Environmental Policy Act (NEPA) which requires the federal government to:

> use all practicable means to assure for all Americans safe, healthful, productive and aesthetically and culturally pleasing surroundings ... and ... develop methods and procedures ... which will ensure that presently unquantified environmental amenities and values may be given appropriate consideration.

NEPA calls for the use of the arts of environmental design, as well as the natural and social sciences, in decision-making, promotes the consideration of environmental values in public decisions, and demands an environmental impact statement on proposed developments. Scenic values and visual impacts are specifically mentioned as major components of environmental quality.

With the passing of NEPA, research efforts were stimulated, public awareness of aesthetic issues improved, and government agencies began slowly to address the issue in the form of training programmes and the production of technical manuals. Despite a severe downturn of environmental interest in the ecologically-ignorant Reagan administration ('when you've seen one redwood you've seen them all'), environmental aesthetics has been successfully enshrined in American public policy.

A mere listing of some of the more important pieces of legislation developed in the period known as 'the Sixties' demonstrates the scale and rapidity of the change:

1965: The Multiple-use Sustained Yield Act (Forestry)
1965: The Highway Beautification Act
1968: The National Wild and Scenic River Systems Act
1968: The National Trails System Act
1968: The National Historic Landmark Preservation Act
1969: The National Environmental Policy Act
1970: The Wilderness Act
1970: The Endangered Species Act

1972:   The Coastal Zone Management Act
1973:   Principles and Standards for Water and Related Resources Planning

The net result of this legislative flurry was to strike a major blow at the traditional American view of 'land as commodity' in favour of a more enlightened, but not yet perfect, view of land as a scenic resource. The use of the word resource is tactically sound, for it imparts an apparent economic value to what may have no tangible economic attributes, and thus neatly conforms to a culture which sees economics as primary.

Britain

National parks came much later to Britain than to the United States, and their character is quite different, reflecting the relative lack of wilderness and the British predilection for the beautiful and picturesque rather than the sublime. Above all, British national parks are countrysides in which people live and work.

Effective public policy on environmental aesthetics in Britain came with the Town and Country Planning Act of 1947, followed by a rash of related legislation, including the National Parks and Access to the Countryside Act (1949), the Countryside Act (1968) and the Civic Amenities Act (1967). Such is the hold of country matters on the psyche of this urban nation that national parks have been complemented by the designation of a whole series of related features, including Areas of Outstanding Natural Beauty, Sites of Special Scientific Interest, National Nature Reserves, and Heritage Coasts (Figure 4.10). An amazing variety of agencies is involved in the preservation and conservation of the English countryside, the more notable ones being the Nature Conservancy, the National Trust, English Heritage, the Ramblers' Association, the Youth Hostels Association, the Cyclists' Touring Club, the Countryside Commission, the Country Landowners' Association, and the Council for the Protection of Rural England. A similar list for urban areas would include such bodies as the Council for British Archaeology, the Royal Town Planning Institute, the Royal Institute of British Architects, and the Civic Trust. The efforts of these and other groups ensured that as early as 1971 almost half the total land area of England and Wales enjoyed some form of protected status (Patmore 1971).

Urban heritage issues and country matters are avidly discussed on national television programmes and regular 'countryside columns' appear in the broadsheet newspapers. Environmentalist groups lobby against the destruction of hedgerows by modern farming practices, which have produced decidedly unBritish 'prairie landscapes', urge the planting of deciduous trees in irregular formations rather than the stark coniferous rectangles formerly favoured by the Forestry Commission, preserve and spread the wildflowers and small animals whose habitats in hedgerows and ponds are destroyed by modern agriculture, and try to redirect huge 400 kV electricity transmission lines from attractive landscapes.

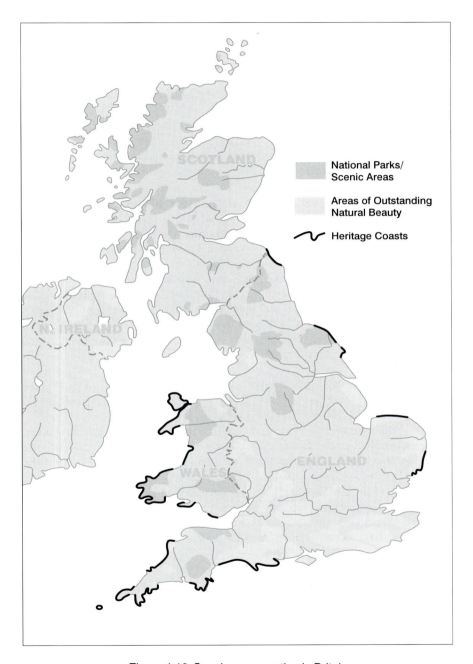

Figure 4.10 Scenic conservation in Britain

Perhaps the most obvious complement to the previous discussion of American wilderness preservation, however, is the British concern for the preservation and conservation of townscapes on a large scale (Lowenthal 1980, 1985, Prince 1981).

The historic preservation, or heritage, movement laid the groundwork. In Britain the preservationist attitude has a long history. Attachment to tangible relics from the past emerged in the Renaissance, and by the late eighteenth century people had begun to see their personal lives as part of the flow of history (Lowenthal 1980). Hence the public (historical) and private (nostalgic) senses of the past came together so that relics and remains were seen as beautiful and meaningful, with a wealth of symbolic values and associations. This movement reached its apogee in the late nineteenth century with the craft medievalism of William Morris and the aesthetic activism of John Ruskin.

Until Ruskin, the obvious way to deal with crumbling ancient buildings had been to restore them. In 1849 Ruskin denounced restoration as 'a lie from beginning to end', and by 1855 The Society of Antiquaries had pronounced that 'no restoration should ever be attempted otherwise than ... in the sense of preservation from further injuries'. In the same spirit, the feeling that no modern had the right to change historic buildings in an inauthentic manner, Morris founded the Society for the Preservation of Ancient Buildings in 1877. Although its initial mission was to preserve medieval buildings, its remit gradually grew so that by the 1930s it had been progressively extended to cover Tudor, Stuart, Queen Anne, Georgian and Regency buildings. By the late twentieth century the preservation movement has again been extended to include Victorian, Art Deco and even some post-World War II early Modernist examples.

Since 1882 Britain has generated a large number of historic building and ancient monument acts. Governments have the power to survey, 'schedule', purchase and preserve buildings of national or local interest. Scheduling includes the making of national lists of all buildings worthy of preservation; the number of 'listed buildings' now runs into hundreds of thousands. The sheer scale of the preservation movement may be seen from its failures; since the 1940s British Rail has pulled down over 200 railway stations, hundreds of churches have been declared redundant since ecclesiastical reorganization in 1968, and by the late 1970s it was estimated that over 800 country houses had been demolished since the beginning of the twentieth century.

Strict preservationists generally abhor conversion or alteration, although they do permit repair and maintenance. Faced with a building about to crumble, such as one of the many medieval cathedrals or later country mansions, the milder preservationist has several options:

- Let it fall into ruin: acceptable to eighteenth-century taste, but not now;
- Modify by removing structurally unsound parts: acceptable in the nineteenth century, but not today because of the concept of 'integrity';
- Conversion, e.g. of churches into homes, arts centres, or offices: an

increasingly viable option in modern Britain;

- Restore: difficult, costly, and aesthetically problematic in that the restoration must be neither obtrusively modern nor obviously falsely antique;
- Demolition: often a viable solution for an owner unable otherwise to develop the site, hence listed buildings may receive so little maintenance that they collapse and thus effectively delist themselves.

Because of the outcries of owners trapped as 'museum curators' in the listed buildings they own, the recent trend has been to promote restoration, modification and conversion, especially of interiors. Thus a genuinely medieval or Georgian street façade may now be found housing an ultramodern high-technology office complex or a block of luxury apartments.

Preservation, therefore, has gradually given ground to the conservation movement. Lowenthal and Binney (1981), in *Our Past Before Us: Why Do We Save It?*, startled the preservationists by asking several pertinent questions, such as: what should we preserve from the past?; how shall we use what we save?; how may we avoid being overwhelmed by a fake past, exemplified by the American practice of creating museum villages from unrelated buildings transported piecemeal to any suitable site?; and how can preservation encourage the creative, practical use of the past? (Figures 4.11, 4.12). The goals of Lowenthal and Binney were preservation at a large scale (whole villages, town centres, streets, and historic landscapes), to promote viable economic usage of historic buildings and landscapes, and to avoid disneyfication and museumization in favour of creative uses – a thatched medieval office at a filling-station, a church converted into a senior citizens'

Figure 4.11 A preservationist problem: Beverley, Yorkshire

club, a country mansion turned into the headquarters of a business corporation.

Preservationism has been strongly attacked by conservationists (Aldous 1972). The latter see preservationists as negative, protectionist, defensive, and concerned far too often with the individual building rather than its landscape or streetscape context (Figures 4.13, 4.14). Since the late 1950s, however, the preservation of individual buildings has been seen as poor policy in the light of the large-scale redevelopment of British cities. There is no point in preserving a single Georgian town house if all its neighbours are demolished in favour of a supermarket or tower block. The new logic, therefore, is to preserve whole townscapes, including those which might not include any 'listed buildings'. This was certainly necessary in the early days of the conservation movement, for many Victorian buildings were not listed prior to the vigorous lobbying of the Victorian Society, founded only in 1958.

Under the Civic Amenities Act of 1967 the concept of the Conservation Area was born. Conservation areas were to be designated by local authorities not because of a few special listed buildings but because of the special character of the whole landscape. The main aim is not merely to preserve but to ensure that change is sympathetic to existing townscape character both in architectural (façades, styles) and planning (usage, traffic) terms. Aldous (1972) provides examples of how many local civic groups have been able to prevent local planning authorities from giving permission for the erection of large tower blocks which may have given a hundred people wonderful views but which would have utterly spoiled the contextual scale of the townscape for the citizen at large.

The idea began only with the 1967 Act, and yet such was the scale of the response that by June 1970 the Civic Trust was able to announce the designation of the one thousandth conservation area. Only a year later the number had reached 1,500. Conservation areas vary greatly in size and character, for they are designated locally and thus are in tune with what Alexander Pope called 'the genius of the place'. They include streets, squares, village greens and town centres, range from 'two houses and the surrounding landscape' to the 800 acre Hampstead Garden Suburb, and generate action ranging from tidying-up through tree planting, the laying underground of wires and collective façade face-lifting, to the limitation or exclusion of traffic to form pedestrian precincts. The movement was given a significant boost by the royal Silver Jubilee celebrations of 1978, and continues to be popular in a nation deeply devoted to the countryside and to the past.

Conclusion

Similar trends in public policy have occurred in most developed nations, and in certain regions of the South. The Swiss land management law of 1979, for example, established zones of protection for areas distinguished by their beauty (Ligue Suisse 1979, Pellegrini 1991). Nevertheless, implementation is often tardy; for example, the new Canadian Environmental Assessment Act

Figure 4.12 Reflections on preservation: Hull, Yorkshire

Figure 4.13 Solitary heritage building in a modernist streetscape (above) with later post-modernist addition (below): Victoria, BC

Figure 4.14 Restoration of a streetscape, Victoria, BC (courtesy Michael Kluckner)

was not passed until 1992, and even by late 1994 had not been implemented.

Moreover, a recent review of a wide range of environmental legislation of the last two decades (Wilson 1991) suggests that one underlying motive behind these policies has been the creation of what are in effect state subsidies for private profit-making on the part of the mass tourism industry and its adjuncts, which range from vehicles to cameras and colour film. Academics have, of course, made their contribution to this development through the boom in largely business-oriented leisure and recreation studies. Yet Wilson notes that state legislation generally fails to address changes in public attitudes, which in recent years have taken a non-consumptive turn which provides some hope for the future (Figure 4.15).

One of the more hopeful recent signs of improvement, which may augur well for improved future legislation, is the growth of environmental advocacy by respected or at least well-known, public figures. Despite good work on public television, most American programming on environment themes is still dominated by majestic landscapes, killer animals, and 'uncharted regions full of bounty' (Wilson 1991: 147), what I call the 'Killers of the Rupununi syndrome'. It is therefore refreshing to find the actress Meryl Streep narrating an Audubon film *Arctic Refuge: Vanishing Wilderness* (1989) which candidly examines the debate about further development on the Alaskan North Slope.

Canadians, however, with their strong tradition of documentary film-making, have best articulated the new environmental advocacy. Unlike typical American porno-nature programming, popular television series such

Figure 4.15 At Stonehenge public enthusiasm conflicted with site preservation: access now restricted

Figure 4.16 Forest clearcutting in British Columbia, an aesthetic and ecological disaster

as *The Nature of Things* examine the full range of social, political, ethical and spiritual aspects of nature and human history (Figure 4.16). The 1985 Canadian Broadcasting Corporation mini-series *A Planet for the Taking*, produced by formerly hard-nosed scientist David Suzuki, provides an accessible yet profoundly philosophical reflection on the Western mission to dominate nature, which it completely rejects. This existing Canadian work adds weight to Malcolm Lowry's (1970: 202) prophetic belief that 'some final wisdom would arise out of Canada, that would save not only Canada herself but perhaps the world'.

Saving the world through advocacy and public policy-making, however, must eventually end either in some form of implementation through planning, or in such a transformation of human consciousness that formal planning becomes unnecessary. As the latter is unlikely in the short term, I turn to aesthetic planning issues in the final chapter.

5

PLANNERS

Public housing, Hong Kong

How to keep – is there any any, is
         there none such, nowhere known some,
         bow or brooch or braid or brace, lace,
         latch or catch or key to keep
Back beauty, keep it, beauty, beauty,
         back … from vanishing away?
                              Gerard Manley Hopkins

During the late twentieth century the majority of the world's population became urban, and a proportion of these also became more affluent. Cities have expanded outwards, upwards, and downwards, and are in a constant state of flux. The corresponding pressures on rural and wilderness areas beyond the city have become immense. Not only do cities physically devour countryside, but growing recreational demands coupled with increased personal mobility ensure that both countryside and wilderness are regarded increasingly as playgrounds for urbanites. Further, the impact of highly-capitalized and mechanized agriculture has significantly altered rural land-scapes, while modernist and post-modernist architectural styles have transformed the centres of most major cities.

Because of these trends, with their inherent problems of land-use conflict, oppositional values, loss of identity, and resource allocation, Western nations have enacted an unprecedented flood of regulatory and conservatory legislation designed to protect and enhance the appearance and integrity of both urban and nonurban landscapes (see Chapter 4.4 above).

Such legislation demands rational implementation on the part of planning agencies, which in turn depend upon the concepts generated by humanists (Chapter 2), rely upon techniques developed by experimentalists (Chapter 3), and are pressured by activists (Chapter 4).

Aesthetic landscape planning falls almost too neatly into two very obvious components, the urban and the nonurban. The second, which I will for convenience term 'landscape', is the province of geographers, foresters, resource managers, highway engineers, and similar professional groups; they are concerned almost exclusively with rural and wilderness landscapes and they have a strong bent toward the quantitative. In sharp contrast, the urban scene, or 'townscape', attracts architects, landscape architects, urban design-ers and planners, and geographers, many of whom take a more qualitative stance.

Except for geographers, whose holistic discipline is all-embracing, land-scape and townscape practitioners rarely meet and only infrequently exchange ideas. It's a pity, for the quantitative/qualitative dichotomy, as elsewhere, bedevils not only theory generation but also the practice of creating aesthetically pleasant, liveable environments. Only in the field of heritage conservation, whether of historic townscapes or historic landscapes, do the two come together conceptually.

Yet their ultimate goal is one and the same. By the 1960s many Western societies were beginning to query capitalism's previously unquestioned growth ethic, and had begun to debate in earnest the compatibility of rapid technological growth with the retention or creation of aesthetically high quality environments. With this in mind, Sylvia Crowe (1971) defined landscape planning as 'planning for the conservation, adaptation and creation of landscapes which will function both with efficiency and beauty, and endure in a sound state of ecological balance'. Although Crowe was chiefly thinking of planning in rural Britain, her statement can be extended to cover all societies and both urban and nonurban environments. Environment is the stage on which human activity is set. Environmental planning is a comprehensive, total planning process which nonetheless pays considerable attention to intangibles, including aesthetics.

In practice, however, urban and nonurban aesthetic planning remain quite separate areas, mirroring, perhaps, the preference dichotomy outlined above (Chapter 3.3). This chapter therefore has separate landscape (Chapter 5.1) and townscape (5.2) planning sections. Whereas townscapers rely fairly heavily on the traditional planning approach of expert evaluation, tempered by public input in a variety of forms, this *modus operandi* is only one of two major thrusts in nonurban landscape assessment.

This concern with planning, of course, is only a short-term panacea. As activists suggest, effective long-term aesthetic improvement of the environment demands widespread and thorough environmental education (5.3). And we must permit both humanists and radicals to criticize the whole enterprise (5.4).

5.1 AESTHETIC LANDSCAPE PLANNING

After the Second World War, the concept of landscape as a resource, and often a scenic resource, began to gain ground in the Western world. The vast and rapid expansion of suburbia provoked fears that rural landscapes would be obliterated by a tidal wave of homogeneous housing developments. In Atlantic Europe, the simultaneous rapid development of a highly-mechanized agriculture, often spoken of as the Americanization of European agrarian practice, led to the disappearance of significant areas of the much-loved *bocage* landscape of small hedged fields in favour of 'prairies'. According to Tinker (1974: 722), 'the traditional scenery of England is fast disappearing, and is being replaced by a countryside with severely diminished visual and wildlife interest. . . . Within a generation, much that we value in the English landscape will be gone' (see also Chapter 4.1, 4.4).

In North America, equally urgently, the growth of leisure time, affluence and automobile ownership had begun to threaten the character of wilderness areas as camping and hiking became popular activities. This widespread perceived threat to rural and wilderness landscapes led to the plethora of protective legislation described above (Chapter 4.4). For example, the United States Water Resources Council's 1970 'Standards for Planning Water and Land Resources' already stipulated that water and land-use plans be evaluated according to their positive and negative impact on environmental quality as well as in terms of economic, social, and regional development. By the early 1970s, Zube (1973: 130) was able to state categorically that 'scenery has become a resource', and, according to Craik (1972a, 1972b: 264) 'systematic landscape appraisal appears destined to form a basis for important environmental decisions'.

Legislation, however, requires implementation via practical planning methods. Those who wish to preserve or enhance landscapes must not only be able to explain why and how; in an age hooked on a narrow definition of science and technology this explanation has to be numerical. Further, given the capitalist economic pressure to 'develop' almost all landscapes for profit, preservationist planners had to be able to demonstrate, using numerical data, which landscapes were most in need of preservation, and which, having less aesthetic, historic, or scientific value, might be regarded as expendable. As a result, a rash of landscape evaluation techniques ensued.

Techniques for general 'environmental planning', 'landscape resource analysis', or 'landscape appraisal' (confusing terminology is a hallmark of the genre) burgeoned in the 1960s. By the end of the decade, Steinitz (1970) was able to identify and evaluate fifteen distinct methods which represented 'the state of the art of landscape analysis'. Most of these used political areas or watersheds as base areas, their goals being to balance the competing demands for land of developers, official agencies, and the public. Computer-generated maps of such features as 'soils suitable for septic tank operation', 'attractiveness for camping', 'aesthetic attractiveness', 'imminence of urban encroachment' and the like were prepared, and land-use competition problems solved by the age-old geographer's map overlay method popularized by McHarg (1969).

Aesthetics, however, generally proved to be but a minor component for those who chiefly looked upon landscape as an economic resource which might have social and political overtones (McAllister 1982). A major problem in highlighting aesthetic considerations was the difficulties encountered in finding convincing methods of assessing landscape aesthetics mathematically. In the 1970s, there was little doubt that we should

> put numbers on values once dismissed as 'just aesthetics.' ... The familiar, vague and subjective feelings about beauty, expressed with emotion alone, are not enough ... quantitative assessment techniques ... will ultimately provide that needed scientific rigour on which a new era of planning and design can expand.
>
> (Fabos 1974: 165)

On both sides of the Atlantic it was felt that unless landscape advocates could present hard data at official land development hearings, their often emotional appeals (see Chapter 4.1, 4.2) would continue to be ignored in favour of the slick presentations of corporations and government agencies armed with charts, graphs, tables, statistics, cost-benefit ratios and other persuasive quantified matter. In response to this need, numerous quantitative aesthetic landscape appraisal methods were developed. As noted by Gold (1980), they fall roughly into two categories, those chiefly involving experts (landscape evaluation methods) and those which derive some or most of their data from the public (landscape preference methods).

Landscape evaluation appraisals

The foundational premise of the evaluation approach is that trained personnel are able to conduct valuable, usable, and publicly-acceptable aesthetic landscape assessments without resort to significant input on the part of the general public or the users of the landscape in question. Evaluation is therefore conducted, independently of the attitudes and preferences of the public, by professional planners and consultants. They attempt to appraise first the significant landscape components and then the 'aggregate landscape' made up of those components, so that sections of landscape may be labelled 'good', 'bad', or 'indifferent' aesthetically, thus providing important input into planning and development decision-making. Essentially, the method involves defining a number of factors which are likely to influence landscape quality variations, establishing scales for measuring these factors, and developing a weighting system to assign varying emphases to the different factors (Dunn 1974a, b, 1976). There are two chief types of landscape evaluation, oddly known as the 'non-evaluative' and the 'evaluative'. A third type is known as the 'expert method'.

Non-evaluative methods

Non-evaluative methods attempt to eradicate the possibility of any value-judgement on the part of the landscape appraiser. They are very few in number and are closely associated with the pioneering work of Luna B. Leopold in the United States. In the late 1960s, Leopold (1969a, b) began to develop 'a method that would quantify the aesthetic features of the environment so that the resultant data could be used in many planning and decision-making contexts'. His basic concept was that it was possible to achieve an objective description of an area's aesthetic uniqueness, which would be quantified in terms of the area's physical terrain factors, biological and water factors, and human use and interest factors. Using these factor types in his initial study of sixteen river valleys, Leopold chose forty-six criteria to describe the aesthetic character of the valleys. Many of these were precisely measurable, including river width, depth, velocity, bedslope erosion, water character, turbidity, presence of algae, trash and litter, length of vistas, degree of view confinement, level of urbanization, and so on. For less measurable features a number from one to five could be assigned.

Intercomparison of sites according to these criteria enabled Leopold (1969b: 40) to produce 'uniqueness ratios' which were 'an objective measure of how different each site is from other sites studied, without regard to "positive" or "negative" aesthetic values'. Heavy pollution could thus render a river section unique. Selection of certain criteria, however, did enable the appraiser to point out which river sections were aesthetically pleasing.

Leopold's aim was 'to eliminate personal subjectivity in landscape analysis' (p. 40). His resulting graphs demonstrated that high quality sites tended to have large rapidly-flowing rivers in large-scale landscapes, with scenic vistas and little urbanization (e.g. the Colorado River in the Grand Canyon). This is no more than we might have expected from our knowledge of American culture (Chapter 2.4). Further, the goal of objectivity is flawed by the eventual need for the appraiser, in phase E of the method, to select which factors are to be deemed most important in terms of high aesthetic value. This fatal blow to the hopeless search for pure objectivity also bedevilled European colleagues who, however, were chiefly concerned with the development of 'evaluative' rather than 'non-evaluative' methods.

The expert method

In sharp contrast to both, the 'expert school' of evaluative practitioners, exemplified by the work of Burton Litton (1974, 1982), accepts the inevitability of subjectivity on the part of the professional landscape appraiser. Litton supports his rather personalist approach, 'apt to be looked upon as subjective, reflective of personal values, and constrained by my own knowledge and experience' (1982: 98) by appealing to that very experience. Like Leopold, Litton asserts the importance of field observation rather than office-based map evaluation, and provides a set of aesthetic landscape evaluation criteria which include judgements of unity, variety, vividness and integrity (concepts normally associated with art criticism; see Chapter 1.1, 1.2) as well as the usual landscape elements of scale, colour, geometric form, pattern density, nature of vegetation edges, and the like (see Chapter 3.2).

Litton believes that 'complete avoidance of personal bias in scenic analysis is probably not possible, but it is a clear goal to seek' (1982: 99). He seeks it via the use of standardized inventories (like Leopold's check-lists), faithful field recording of visual elements and relationships, sympathy with known public perceptions, and by working closely with residents of the landscape being appraised. His work is essentially qualitative, based on his assertion that 'through the long history of cultural development, aesthetics as the search for beauty has not expressed urgency for the development of a quantification system. Expecting precise measurement of landscape quality is best accepted as an unsolvable problem' (1982: 104). In this he resembles the traditional school of urban aesthetic appraisal (see Chapter 5.2). But although Litton's work was a major influence in the United States into the 1980s, the spirit of the age demanded the relentless application of quantification in the pursuit of the will-o'-the-wisp known as objectivity. Thus we now turn to the second major appraisal method, the 'evaluative' approach.

Evaluative methods

Although not the first (Fines 1968), one of the best-known early attempts at evaluative aesthetic analysis of landscape was that of David Linton (1968) in Scotland. More reductive than Leopold, Linton saw only two 'truly basic' elements in the scenic resources of an area. These were, first, the terrain's geomorphology and, second, the overlying mantle of vegetation and human artifacts. Six 'land-form landscapes' typified the former, and were awarded points as follows: lowland (0), low upland (1), upland plateaux (3), hill country (5), bold hills (6), and mountains (8). Seven land-use features were similarly awarded points: wild areas (+6), richly varied farmland (+5), varied forest and moorland (+4), moorland (+3), treeless farmland (+1), continuous forests (−1) and urban land-uses (−5). By adding together the land-form and land-use points for any area, argued Linton, the appraiser could produce a composite scenic resources map with values ranging from +14 (wild mountains) to −5 (urbanized lowlands). The result for Scotland is illustrated in Figure 5.1, and according to Linton (p. 238) 'This assessment is objective.'

Although the aesthetic map of Scotland demonstrates that Linton was roughly in tune with what many readers might expect (given the Humanist analysis above, Chapter 2), as an exercise in objectivity Linton's method is woefully inadequate. Linton, a distinguished geomorphologist, loved mountain scenery. Although he recognized that cities might contain 'whole townscapes of great charm' (p. 231), he refused to accept these as 'scenic resources' and declared that 'the scenic resources would be greater if they

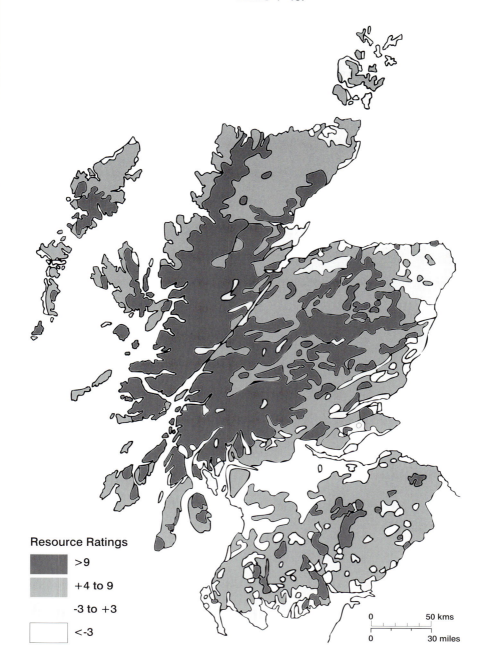

Figure 5.1 The scenic resources of Scotland (simplified from Linton 1968)

were not there'. This is an amazing rejection of city aesthetics (Chapters 2.3, 5.2)! The graceful streets of Edinburgh's eighteenth-century planned town-scape are accorded the same negative rating as the ugly wastelands of the Clyde Valley.

Similarly, Linton had little time for forests (popular in Germany and North America) and paid no attention to the subtleties of lowland scenery. His insistence that his method had universal application betrays ignorance of gross cultural differences (Chapter 2.2, 2.4) and, *inter alia*, relegates his neighbours in Humberhead, the Fens, and the Netherlands to an aesthetic sink-hole; so much for Dutch genre landscape painting. Charges of methodo-logical slackness, unexamined assumptions about the nature of aesthetic experience, and sheer unadulterated bias abound in the literature (Wright 1974, Appleton 1980).

Yet Linton's work was refined and extended by Gilg (1974a), who subjected the method to tests in Scotland and elsewhere. He concluded that the technique produced useful results, but suffered badly from an over-emphasis on relief, the inflexible condemnation of urban landscapes, and lack of definition of 'a view'. But with suitable modifications the technique was found to produce consistent results, an attribute not always inherent in other landscape evaluation models (Brancher 1969, Penning-Rowsell and Hardy 1973, Gilg 1974b). The greatest asset, however, is that the method produces 'valid results quickly and with relatively untrained personnel' (Gilg 1976: 41). In other words, it is a 'quick and dirty' method for defining some of the basics of aesthetic landscape quality, but 'cannot be refined into a planning tool for day-to-day management decisions' (Gilg 1974a: 129).

Nevertheless, the pioneering work of Fines and Linton proved inspira-tional, and a large number of similar methods were produced by academics, landscape consultants, and government agencies. For example, the Coventry–Solihull–Warwickshire Sub-Regional Planning Study Group (1971) used a kilometre-grid base to provide a regional aesthetic inventory derived from measurements of land-form (e.g. shape, slope), land-use (e.g. residential, rural), and land features (e.g. 'natural', 'man-made'). From this were generated two major indices, one of landscape value based on visual quality, the other of intervisibility, based on views between grid-squares. By combining these two indices, a map of 'landscape value' was produced (Fig. 5.2).

A much more thorough attempt at objective evaluation became known as the Manchester Landscape Evaluation Method (Robinson *et al.* 1976). This Manchester University research team produced scores of working papers, one of which reviewed several existing techniques and found them all wanting except those of Fines (1968), which required 'a vast expenditure of resources', and Coventry–Solihull–Warwickshire, which unfortunately failed to provide objective measures of the crucial visual quality components (Traill n.d.). In turn, the Manchester method proved too statistically complex to be readily usable in planning offices and, moreover, relied for its evaluation data on the knowledgable but nonetheless subjective judgements of a small

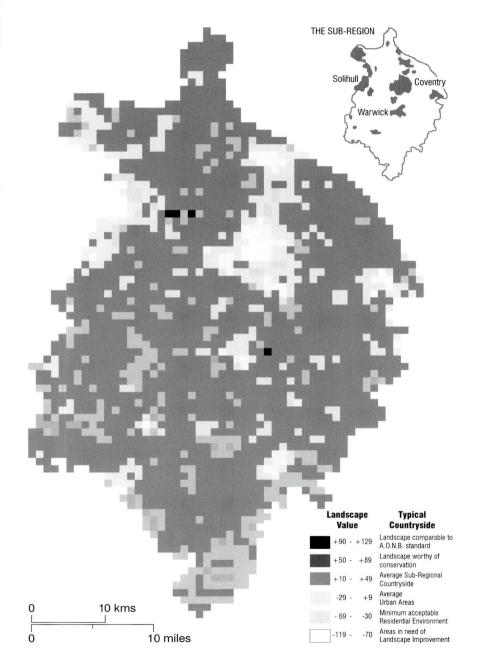

Figure 5.2 Landscape values of the Coventry–Solihull–Warwickshire region,
England (after Penning-Rowsell 1974)

corps of 'mature and experienced members of the environment and design professions', whose views may or may not coincide with those of the public (Penning-Rowsell and Searle 1977). In response, the Manchester team asserted that theirs was one of 'the few current methods that is both workable and theoretically sound', but also conceded that 'the statistical approach to landscape remains very much in its infancy ... any attempt to apply mathematics to aesthetics will be met with total incredulity in many quarters' (Traill 1978: 3).

The burgeoning number of techniques was soon subject to comparative analysis (Crofts and Cooke 1974, Crofts 1975, Liddle 1976). The chief consensus to emerge was that all these much-vaunted objective techniques suffered from a single major flaw; fundamentally, they were all too subjective!

Liddle (1976), for example, compared numerous methods according to their techniques for: selecting landscape features to be measured; measuring these components; and analysing the measurements. Although a high level of objectivity was possible in the latter two stages, 'the subjective choice of features to measure must remain the weakest link unless those that have primary importance in our perception of landscape can be defined' (p. 181); any such definition, of course, would be to some degree subjective. Similarly, Crofts (1975) reviewed a variety of landscape components frequently used as surrogates for landscape attractiveness and came to the conclusion that 'there are so many pitfalls in the techniques described that some critics may regard this type of surrogate approach as being virtually useless; but there are few more satisfactory methods available at the present time' (p. 129).

Given the obvious subjectivity inherent in methods touted as 'objective', critics clearly had a field-day. Not only were the techniques suspect, but the whole enterprise was insufficiently grounded in theory. Craik (1972a), for instance, suggested that the appraisals of landscape quality which emerged from such operations might only poorly predict public preferences for the landscape in question. Turner (1975) noted that most British landscapes were products of aesthetic or economic decisions made centuries ago, and that individuals enjoy best the landscapes they grow up with. Penning-Rowsell (1974) asserted that all such 'objective' measures are irredeemably tainted by the subjective opinions and choices of both the technique's creators and the planners who implement the technique in a particular area. He concluded that it might be better to ask the actual users of a landscape to evaluate its quality: 'if planners want to produce definitive landscape evaluations that truly reflect the landscape attractiveness of their ... districts, they must approach the problem via the user-dependent or "participatory" route, and not get misled up the methodological blind alley of so-called objectivity' (p. 934). Which brings us neatly to the second major type of aesthetic assessment of landscape, the landscape preference study.

Landscape preference appraisals

The fundamental differences between the evaluative and preference approaches are, first, that the latter go directly to the public, the landscape's users, for data, and second, that they generally require a judgement of landscape in its totality, 'rather than as the sum of a number of component parts' (Dunn 1976: 16). As Fines (1968) notes: 'The value of a landscape composition is certainly greater than the aggregate value of its component parts.'

The case for adopting user-preference methods rather than evaluative studies is rather strong: to find out what the public, who ostensibly are those who are being planned for, prefer in terms of landscape aesthetics, planners should either ask them directly or at least observe their behaviour. Two basic preference-based approaches therefore emerge.

Behaviour-observation methods are rare; they assume that behaviour mirrors preference, a rather questionable assumption. Weddle and Pickard's (1969) method includes a section which attempts to assess 'acquired landscape value'. Visitor counts are made at each location to assess its popularity. When these data are combined with expert-appraised inherent landscape quality figures, a final aggregate landscape value emerges, which rates locations from very high to low in aesthetic terms. Similarly, Price (1976) attempted to relate aesthetic judgements to holiday expenditure and trip distributions. Such market-oriented techniques inevitably suffer from the drawback of attempting to infer mental states from patterns of overt behaviours (Gold 1980: 194), and ignore behavioural constraints.

Far more common are interview-based preference methods which recognize that aesthetic landscape quality is a subjective phenomenon which may best be assessed via techniques derived from psychological testing and sociological surveys. The theory underpinning this approach was reviewed earlier (Chapter 3.2) in some detail. The most prominent pioneering work was performed by Shafer and his associates in the United States (Shafer *et al.* 1969, Shafer and Mietz 1970).

Practically, input from landscape users via questionnaires and similar devices is obtained either directly, in the landscape itself, or indirectly, through the use of surrogates such as photographs or colour slides. Again, the basic research on landscape preferences using surrogates has been outlined in Chapter 3.1 and 3.2, above, while 3.4 contains a lengthy discussion of the methodological problems involved. Briefly, surrogates ignore sensory modalities other than vision, present a selective view which is dependent on photographic skills, and cannot take into consideration all the situational factors (such as recreational activities engaged in by respondents) which are likely to influence those who are expected to produce meaningful numerical responses on observing the surrogate.

Nevertheless, Shafer *et al.* (1969) concluded that 'it seems possible to quantify aesthetics' via the use of surrogates, and their work was followed by a flood of academic critique, experiment and refinement which resulted in a number of techniques being used for the assessment of forest landscapes, scenic highway corridors, and the like (Zube *et al.* 1973, Dunn 1976, Thayer *et al.* 1976, Evans and Wood 1980, Tooby and Cosmides 1990). Some of these methods involved validation in the form of testing on-site versus surrogate-based evaluations. Fairly high levels of similarity generally resulted, supporting the use of the relatively quick and inexpensive surrogate approach. Indeed, Zube, Pitt and Anderson (1974) provide some evidence to support Dunn's (1974a, b) assertion that there is a high correlation between both kinds of preference-data elicitation techniques.

Methods which involve on-site interviews solve most of the problems of using surrogates but introduce others; they are laborious to carry out and the opportunity sample obtained may be representative only of those who visit the site in question, not of the population at large.

Penning-Rowsell's (1974) study of landscape preference in the Wye Valley was an early attempt at on-site preference modelling. The researcher divided the landscape up into different tracts based on topography and land-use, and then asked visitors to rank these on a semantic differential scale ranging from 'extremely attractive' to 'extremely unattractive'. The demographic characteristics of respondents were used to clarify sub-group preferences, and maps of individual, group and general consensus preference appraisals were then prepared (Fig. 5.3). Similar, but much more complex work has been performed in Grindelwald, Switzerland (Grosjean 1986) and in American national parks (Baumgartner 1984).

A more sophisticated approach was taken by Dearden (1980a, b) who first measured one-kilometre grid squares in terms of thirty landscape elements hypothesized to influence visual landscape quality, and then asked a selected team of observers to rigorously evaluate a random sample of grid squares in the field. By using multivariate regression techniques (Dearden and Rosen-blood 1980) it proved possible to predict the visual quality of every grid square in the area as if they had been surveyed on the same basis as the sample. The range of quality values thus produced was then divided into classes and the results mapped to produce a map of visual quality for the whole region. Dearden demonstrates that the production of such a map is not beyond the capabilities of the average planning department, thus disarming those critics who believe that in-field methods are too laborious and expensive.

In sum, Dearden (1981) rejects the existing studies which suggest considerable similarity in evaluative judgements between professional planners and the public (summarized by Zube 1976a), and concludes that until it is fully proven that there are no significant differences between the landscape perceptions of planners and those being planned for, public preference studies should be undertaken on both philosophical and pragmatic grounds.

Appraising the appraisals

When Penning-Rowsell (1973, 1975) compared over forty mainly British evaluation and preference methods in the mid-1970s, he found that

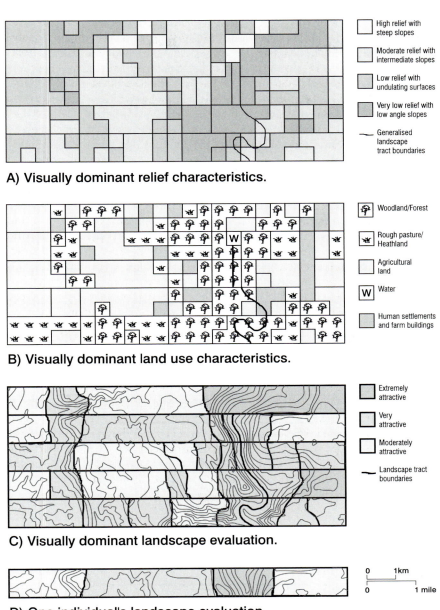

A) Visually dominant relief characteristics.

High relief with
steep slopes

Moderate relief with
intermediate slopes

Low relief with
undulating surfaces

Very low relief with
low angle slopes

Generalised
landscape
tract boundaries

B) Visually dominant land use characteristics.

Woodland/Forest

Rough pasture/
Heathland

Agricultural
land

W Water

Human settlements
and farm buildings

C) Visually dominant landscape evaluation.

Extremely
attractive

Very
attractive

Moderately
attractive

Landscape tract
boundaries

D) One individual's landscape evaluation.

Figure 5.3 Preference evaluation in the Wye Valley, England (after Penning-Rowsell 1974)

evaluation methods accounted for about three-quarters of all methods then in use. Preference methods accounted for less than 15 per cent. Penning-Rowsell (1975: 2) commented that 'there appears to have been [little] development in the "consumer-dependent" techniques dependent on landscape users' preferences and attitudes, perhaps surprising given Planning Departments' continued concern with the participation of the public in their plan-making'. A naive comment, perhaps; elsewhere I have demonstrated that public participation is generally more a stated goal than an accomplished fact (Porteous 1977, 1989).

By the 1980s, aesthetic landscape appraisal methods had become highly technical. This proved especially so in North American coastal zone (Wohlwill 1978, Carls 1979, Smardon and Felleman 1982, Wohlwill 1982) and forest (Daniel *et al.* 1973, 1977, Dearden 1983, Bannerman and Dearden

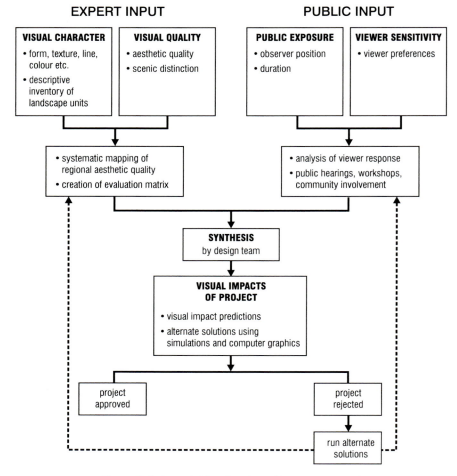

Figure 5.4 A model for visual quality management

1985) management. For example, manuals and models (Figure 5.4) have been developed to aid forest managers in preserving some of the aesthetic appearance of forest landscapes (Figure 5.5) while allowing the maximum permissible cut (e.g. Yeomans 1983; for a dissenting view see Chapter 4.1, 4.2). Their use, however, is rarely fully implemented (Moss and Nickling 1989) and the emphasis remains firmly on utility (Dearden 1983). The peruser of these manuals quickly finds herself awash in a sea of acronyms (another hallmark of the field) such as VQOs (visual quality objectives), PEQIs (perceived environmental quality indices), and VACs (visual absorption capacities), all leading via complex step-wise planning models to efficient VRM (visual resource management), itself only part of TRD (total resource design). The planning tool which has proved of most value in forest management is the VQO.

Yet all through the late 1970s and 1980s a vigorous debate proceeded on the relative merits of what had become a veritable smorgasbord of aesthetic landscape appraisal methods. A review of a large number of these reviews reveals very little guidance for the practitioner who is searching for a suitable method or combination of methods (Dearden 1987). There is considerable consensus, however, on the problems involved in using many of the methods now available.

After a thorough and useful review of the state of the art at the time, Arthur and her colleagues (Arthur and Boster 1976, Arthur *et al.* 1977) suggest that most methods should conform to the following criteria: be based on public experience, rather than expert evaluations alone; involve statistically valid and reliable techniques; be adaptable to a variety of planning situations; and be simple and inexpensive to use. This considerably reduces the field.

At a finer level of analysis, researchers have noted the following problems:

1. The choice of landscape components (or independent variables) for analysis is critical in every method. The choice of landscape dimensions to analyse frequently reflects the unconscious bias of the investigator (Wilson-Hodges 1978). A good example is Linton's (1968) rating of 'richly varied farmland' higher than 'continuous forests', a ranking not likely to be valid in much of North America. Further, the chosen elements often appear to be those most amenable to statistical analysis. Wilson-Hodges (1978: 36–7) asserts that landscapes are multidimensional and, following Litton (1977), provides a long list of relevant variables which are generally excluded as aesthetic landscape appraisal criteria. Clout (1972) similarly emphasizes this subjectivity in what are often regarded as objective methods. Brancher (1969: 91) merely dismisses such methods as 'subjectivity dressed as science'.

2. Research design and statistical techniques have also been critiqued. Wilson-Hodges (1978: 39–41) provides a rather devastating appraisal of respondent sampling procedures, the use of photograph surrogates, and

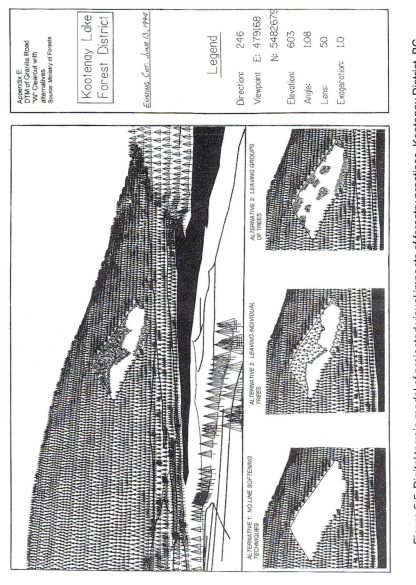

Appendix E:
DTM of Granite Road
"W" Clearcut with
alternatives
Source: Ministry of Forests

Kootenay Lake
Forest District

Existing Cond: June 13, 1994

Legend

Direction: 246
Viewpoint   E: 479168
                  N: 5482679

Elevation: 603
Angle: 1.08
Lens: 50
Exaggeration: 1.0

ALTERNATIVE 1: NO LINE SOFTENING
TECHNIQUES

ALTERNATIVE 2: LEAVING INDIVIDUAL
TREES

ALTERNATIVE 3: LEAVING GROUPS
OF TREES

Figure 5.5 Digital terrain model of projected visual impacts of forestry practices, Kootenay District, BC
(reprinted by permission of the Ministry of Forests, British Columbia)

dubious statistical manipulations, which largely echoes my critique of
the experimentalists (above, Chapter 3.4). Weinstein (1976) suggests that
these statistical problems 'reflect the youthfulness of research on
environmental preferences' (p. 624). Rather more fundamental is the
very general assumption that, once analysed, the sum of the individual
landscape components can be added together to form a picture of the
total landscape (Wilson-Hodges 1978). Similarly, preference means hide
a wide variety of individual differences in personal landscape evaluation.
In either case, the aggregate figures which emerge are often too broad-
brush to be useful (Penning-Rowsell 1981). Nor is there very much
replication or testing of models, a necessary procedure if incremental
improvements are to be achieved (Wilson-Hodges 1978, Penning-
Rowsell 1981a).

Most recently, Hamill (1989) claims that 'there appears to be an
increasing incidence of errors in the literature of landscape aesthetics, not
the decrease one might expect as the result of increasing knowledge' (p.
197). Hamill's typology of errors is worth inspection by anyone
attempting an aesthetic assessment, for it includes accounts of the misuse
of classificatory systems, the use of 'spurious numbers', the generation of
'bad data', the use of concepts without adequate operational definition,
the persistent usage of data which do not satisfy the requirements of the
model, and the 'use of numbers to support, derive, or demonstrate
meaningless, spurious, or useless concepts' (p. 200). Leopold's method is
subjected to especially detailed criticism.

3. Although preference studies appear to be more politically correct than
    evaluations, they are also subject to a wide range of criticisms. Whose
    opinions should be sought? Once obtained, whose opinions should be
    heeded? Harry *et al.* (1969) and Clout (1972) argue that conservation and
    landscape aesthetics are essentially upper-middle-class movements, with
    less appeal to working-class groups. Moreover, it is clear that sub-groups
    and individuals appraise landscapes in very different ways; beaches, for
    example, are evaluated by youths and the elderly according to very
    different criteria (Peterson and Neumann 1969).

    Dearden (1981) and Grosjean (1986) note that there is no single public
    but rather a mosaic of many publics. While Zube and his associates
    (1975) identified as many as eleven sub-groups, Dearden (1981) suggests
    that the major divide is between those with some environmental training
    and those without. Ultimately, the choice is an ideological or political
    one.

4. Equally problematic is the criticism that current methods are not yet of
    great value to planners because they are not readily capable of predicting
    the changes in landscape aesthetics that would be brought about by given
    changes in land-use. Surveying a British plan, Countryside Commission
    reviewers judged that it 'seems to be more a response to existing
    conditions than an attempt to anticipate and provide for future develop-

ments' (Davidson and Tayler 1970: 44). As Jacobs (1975: 149) warns: 'Change is the fundamental prospective issue that must be explored … in the circulation and application of the next stage in the development of models of the landscape image.' Change, or its prevention, of course, is what planning is all about.

In summary, it is clear in the mid-1990s that no solid consensus on the most valuable package of aesthetic landscape quality appraisal methods has yet been achieved. Most existing techniques are refinements of the pioneering methods developed in the 1970s. Few individual methods, moreover, are pure examples of a particular type. For example, Leopold's 'non-evaluative' method demands value judgements in Stage E. Similarly, public preferences are generally given within the constraints of planner-generated evaluative categories and landscape tracts. Most methods, then, are composites of evaluative and preference types, if only because professional preferences are always present in even the most apparently objective technique.

Further, enthusiasm for the creation, refinement and even application of appraisal methods waned somewhat in the destructively market-driven Thatcher–Reagan years of the 1980s, from which Britain and North America are still recovering. Insufficient coordinative effort during this era has resulted in the retention of what Penning-Rowsell (1973: 6) once called 'a startling variety' of appraisal methods, some of them 'flatly contradictory' (Dunn 1974b: 935), and which appear to be used by local and regional planning agencies in such an *ad hoc* manner that inter-regional coordination or comparison generally proves impossible.

For a glimpse of the seemingly endless array of approaches and techniques the reader should consult several volumes, notably those edited by Zube, Brush and Fabos (1975), Appleton (1976), Sadler and Carson (1982), Penning-Rowsell and Lowenthal (1986) and Dearden and Sadler (1989). It will be readily apparent that aesthetic landscape appraisal remains a richly diverse and often highly individualistic activity, which is not necessarily a bad thing.

Recent reviews of practice in Canada, Britain, and the United States are revealing. The situation in Britain seems most dispiriting (Penning-Rowsell 1989). Government-sponsored research designed to produce maps of 'landscape character' rely on remote sensing and professional field survey, and deliberately eschew aesthetic value judgements. 'Grand' landscapes are favoured over everyday ones. A general waning of interest in the subject is apparent, except in public preference research, which unfortunately has little opportunity for application.

The only major positive development is a growing interest among landscape architects and consultants in the practice of preserving, enhancing, and restoring historic landscapes, parks, and gardens. In this connection, the planning process involved in the restoration of the famous park at Blenheim, Oxfordshire is of interest not only for the variety of inputs involved but also for its illustration of the relationships between visual appearance and a host of other significant physical and cultural factors (Moggridge 1984, Fig. 5.6). But controversy abounds even in landscape restoration. At the one extreme are those who would mercilessly root out extensive later growth in order to restore as exactly as possible an example of a Capability Brown or picturesque parkland. For instance, the Hafod Trust is currently restoring an important Welsh picturesque landscape laid out in the late eighteenth century by a cousin of Richard Payne Knight. Critics revile such restorations, believing that eighteenth-century parkland landscaping offers a false representation of nature which has severely warped modern sensibilities. In the case of Kenwood House in London, a compromise has been reached between the purist restorers and those who would retain the present overgrown 'jungle'. Interestingly, a survey indicated that those who visited Kenwood Park most often wanted the least change, while infrequent visitors were most strongly supportive of extensive restoration (Evans 1995).

Public input is clearly important in deciding the look of the landscape. An agenda of important research topics identified by the Landscape Research Group (Penning-Rowsell and Lowenthal 1986, Penning-Rowsell 1989) confirms this embedding of aesthetic considerations in a much wider matrix of landscape research possibilities which includes consideration of public values. Overall, however, Penning-Rowsell feels that 'landscape research in the United Kingdom today appears both disparate and directionless' (p. 243). It is perhaps significant that a recent British landscape-architecture text (Preece 1991) contains a chapter entitled 'Accusations of subjectivity'.

The United States report is more upbeat, but, like Penning-Rowsell, it comments that 'though there has been a growing volume of public preference research, this work has had little influence on applied visual assessment and visual impact studies' (Itami 1989: 231). One reason for this is that landscape-planning professionals lack the research training necessary to operationalize preference methods; more important is the finding that they are apparently quite satisfied with existing evaluative methods such as the Visual Management System in use in the Forest Service (Laughlin 1984). Similar systems are used by the Bureau of Land Management and the Federal Highway Administration (Table 5.1).

Itami notes, however, that there has been very little empirical research on the reliability, validity, generalizability, or efficiency of these systems, a factor of paramount concern to those who prize methodological rigour (see Chapters 1.4, 3.4). In short, says Itami, while the preference school sacrifices generalizability and utility to achieve high standards of reliability and validity, 'the expert paradigm sacrifices validity and reliability for the cause of utility – getting the job done' (p. 225). We are back to the dilemma of Figure 1.4! In this contest utility usually prevails. Itami confirms that a body of widely-accepted visual landscape quality measurement procedures now exists; the challenge is to revise these to include meaningful public participation. Zube (1991: 331) disagrees, for he believes landscape appraisal research 'has had a narrow focus, both topically and methodologically. We know very little about how individuals and groups use these landscapes, about the

SURVEYS, ANALYSES AND PLANS

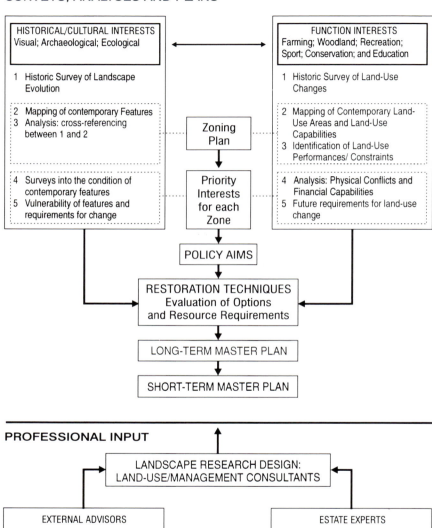

| HISTORICAL/CULTURAL INTERESTS
Visual; Archaeological; Ecological | | FUNCTION INTERESTS
Farming; Woodland; Recreation;
Sport; Conservation; and Education |

1 Historic Survey of Landscape Evolution

1 Historic Survey of Land-Use Changes

2 Mapping of contemporary Features
3 Analysis: cross-referencing between 1 and 2

Zoning Plan

2 Mapping of Contemporary Land-Use Areas and Land-Use Capabilities
3 Identification of Land-Use Performances/ Constraints

4 Surveys into the condition of contemporary features
5 Vulnerability of features and requirements for change

Priority Interests for each Zone

4 Analysis: Physical Conflicts and Financial Capabilities
5 Future requirements for land-use change

POLICY AIMS

RESTORATION TECHNIQUES
Evaluation of Options
and Resource Requirements

LONG-TERM MASTER PLAN

SHORT-TERM MASTER PLAN

PROFESSIONAL INPUT

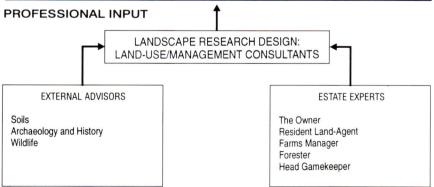

LANDSCAPE RESEARCH DESIGN:
LAND-USE/MANAGEMENT CONSULTANTS

EXTERNAL ADVISORS

Soils
Archaeology and History
Wildlife

ESTATE EXPERTS

The Owner
Resident Land-Agent
Farms Manager
Forester
Head Gamekeeper

Figure 5.6 The process of creating a landscape restoration plan for Blenheim Park, Oxfordshire (after Moggridge in Penning-Rowsell 1989)

Table 5.1 Comparison of US visual assessment and impact systems

	USDA Forest service	USDI Bureau of land management	DOT Federal highway administration
Landscape classification	Physiographic units Character types Subtypes	Physiographic units set landscape context	Physiographic units Viewsheds from highway corridors
Scenic assessment	Visual variety of land-forms Water-forms Rock-forms and vegetation	Land-form scale & ruggedness Vegetation variety Water/size/dominance Colour intensity Adjacent scenery Uniqueness Negative cultural features	Vividness Intactness Unity
Public sensitivity	Importance of use areas Distance zones Viewsheds	User attitudes and use volumes Distance zones Viewsheds	Use volumes Land-use compatibility Local cultural values
Management objectives	Scenic assessment and public sensitivity combined to generate management objectives	Scenic assessment and public sensitivity combined to generate management objectives	Not applicable
Visual impact assessment	Visual absorption capability estimates site screening potential	Visual contrast rating for existing landscape and proposed uses	Photographic simulation

Source: Itami (1989: 218)

meanings they associate with them and about the relative importance of esthetic values compared with the host of other values such as ecologic, historic, economic, and symbolic.' Zube's recent more qualitative stance appears to be close to that expressed contemporaneously in Britain.

As might be expected, Canadian experience (Moss and Nickling 1989) strikes the happy mean between British and American extremes. Practitioners claim that by using modifications of what are essentially Leopold and Linton methods, scenic resource analysis maps can readily be produced for both small (Fig. 5.7) and large (Fig. 5.8) areas. Moreover, work has begun on ways of integrating aesthetic criteria and other relevant landscape components which by their nature are likely to induce visual quality changes, notably geomorphic processes such as gullying, slumping, soil creep and river-bank erosion (Fig. 5.7). The vexed subject of time is thus being addressed for natural changes at least. Moss and Nickling conclude that traditional methods of aesthetic landscape analysis are useful in identifying areas of high scenic value but 'as yet provide little information for environmental planning over a larger time period' (p. 190). Moreover, as yet the demand for scenic assessment data remains 'poor and ill-defined' (p. 190).

All of which should leave us with the same sense of optimistic caution that pervaded the introduction to *Landscape Assessment* (Zube *et al.* 1975) a generation ago. Although 'much of what passes for esthetic evaluation is inadequate' (Arthur, Daniel and Boster 1977: 126), incremental refinements to existing methods continue to be made and are 'crucial to effective public resource management'. Cherulnik (1993), for example, examines a number of cases where researchers have made valuable contributions to scenic resource management and practical policy-making.

It is clear, however, that a single 'good means of measuring the scenic values of landscapes' (Wilson-Hodges 1978: 46) is unlikely to be found. Nor will it be found until some means is discovered to overcome the methodological and political problems which prevent the successful marrying of evaluation and preference methods. In this context, we should also heed the words of Dearden (1982: 238):

All landscapes are not equal. There is no one *correct* way of gauging landscape value. In different locales and at different times under different circumstances one methodological approach may prove more fruitful than another. As students of landscape, we should be sufficiently sensitive to these variations and adjust our methods of approach accordingly.

Which does not deny the need for further advances in research and application, be these philosophical, theoretical, contextual, methodological, institutional, professional or political (Sadler and Dearden 1989). The aesthetic appraisal of landscapes is now entrenched in the planning process and, despite its many methodological shortcomings, is likely to remain so.

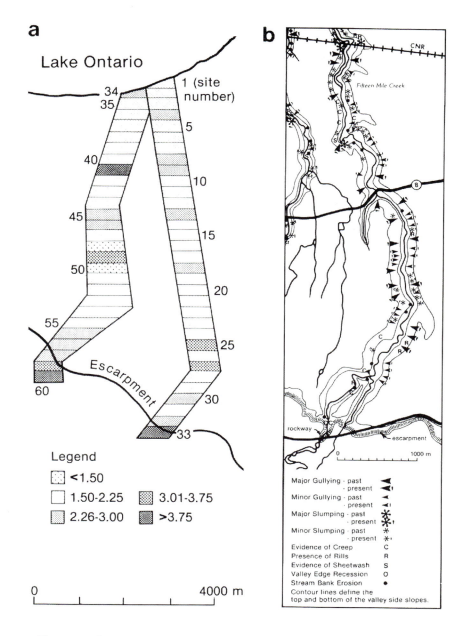

Figure 5.7 Scenic resource analysis of two creeks in the Niagara peninsula, Ontario: (a) uniqueness ratios using the Leopold method; (b) hazard features likely to induce visual quality changes (after Moss and Nickling 1989)

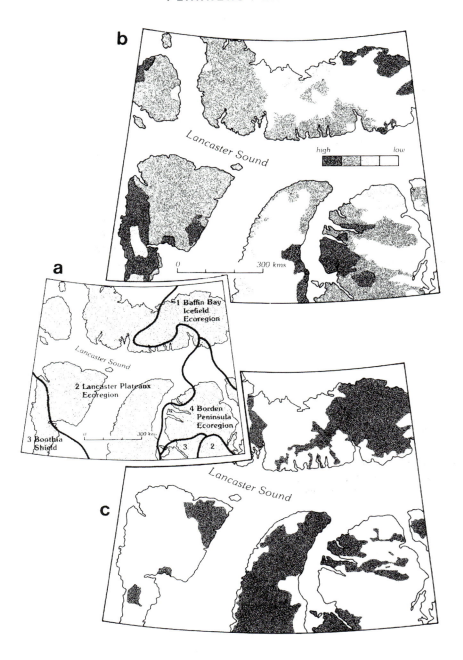

Figure 5.8 Scenic assessment of the Lancaster Sound region, Canadian Arctic: (a) the basic ecoregions; (b) relative scenic value; (c) unique areas (after Moss and Nickling 1989)

5.2 URBAN AESTHETIC PLANNING

There is no shortage of classic or contemporary works on the historical development of townscapes, human behaviour in cities, urban landscape critique, or urban design theory. The latter, *inter alia*, attempts to understand the motives behind the twentieth-century movements that have led to the centres of most Western cities being dominated by first Modernist and then Post-modernist architecture (Chapter 2.3). Lang (1991) notes that the salient fact about Modernist architects is that, lacking the environment and behaviour research that developed after 1950 (Porteous 1977), they relied heavily on common sense, intuition, and ideology. Modernist theories were based, first, on an often-socialist sociopolitical platform for the reorganization of society, and second, on an aesthetic theory that stressed honesty and simplicity to the point of brutality. The results can be seen in Moscow as well as in New York, and a brutally honest central forest of cubes had become associated, in the Western mind at least, with true 'cityness' (Fig. 5.9). Berleant (1992: 92) comments that much city form outside downtowns, however, is not designed by well-known architects, with the result that 'we live surrounded by surfaces, not content, environments that provide images, not substance. Behind this tawdry veneer lie accretions of dull boxes formed without imagination in the name of practicality.'

Like landscape appraisers of the evaluative persuasion, most urban designers have simply assumed that public preferences would coincide with the designers' own predilections. Nay more, for Modernist architects, at least, felt that the moral absolutism and physical perfection of their designs demanded acceptance by the public and adherence on the part of the design professions in general. 'Based on their own observations and values they designed from the top – the intelligentsia – to the bottom, the vulgar people. Not much has changed' (Lang 1991: 79).

This elite-based normative architectural culture was challenged from the 1960s by the Post-modern movement, whose very name highlights either uncertainty about its identity, its embrace of multiplicity, or, more probably, both. Post-modernism is populist without admitting public input, but at least Post-modernists recognize that the concept of a unified 'public' is a myth and that individual and group differences in aesthetic preference are important (Foster 1985). The chief theorist, Christopher Alexander (Alexander *et al*. 1977) bases his designs on human behavioural needs, but finally takes a normative stance, and almost all the contemporary 'Post-modernist heroes' (Dixon 1988) have developed normative positions without much reference to public preferences or desires. Rival contemporary movements, such as Deconstructionism (Jencks 1988, Cornwell 1989) or the latest Classical Revival à la Prince Charles (see Chapter 4.4) are no better. In architecture, as in landscape aesthetic practice, the professional continues to know best.

Perhaps because of this extremely elitist stand, urban designers, unlike evaluative landscape appraisers, have developed very little in the way of procedural methods which could be subjected to critique and refinement by

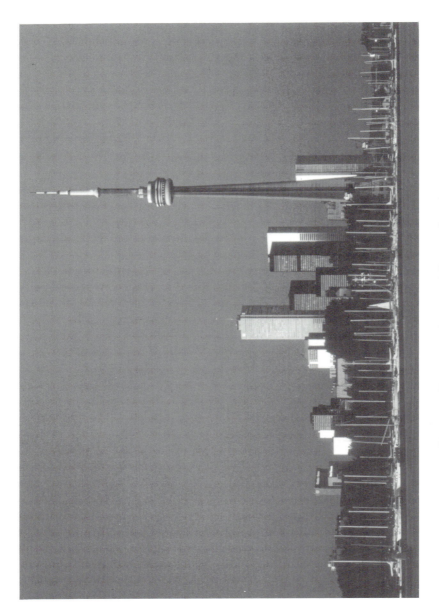

Figure 5.9 The image of 'cityness': Toronto, Ontario

those outside the profession. Architects still proceed like demigods working in an Olympian closed shop. The little we know about the design process still derives largely from the classics, Vitruvius, Alberti, and Palladio, from whence emerges Le Corbusier's (1923) model: formulate the architectural problem in terms of the activities to be housed; formulate design standards, including aesthetics; and intuitively integrate these into an architectural work of art. Despite a flurry of design process studies since the 1960s (Jones and Thornley 1963, Moore 1970, Koberg and Bagnall 1974, Zeisel 1981, Rowe 1987) 'the process remains obscure and there is still much that we do not understand' (Lang 1991: 79).

In the contemporary Post-modernist scene, for example, aesthetic themes differ remarkably: 'Rossi is historical, Krier neoclassical, Leon romantic-Marxist, Ungers rationalist' (Jencks 1988, Lang 1991: 88). Others see the urban environment chiefly as a collage reflecting a medley of architectural attitudes. Few bother to think of what it might be like to live in such a collage.

All of which points up what must by now be glaringly apparent to the reader, namely the profound dichotomy between urban and nonurban aesthetic appraisal and design. Given the general applicability of landscape preference methods (Dearden 1981), it seems remarkable that so little use has been made of them in urban areas. This stems in part from the different training backgrounds of the professionals involved. Whereas rural and wilderness appraisers enjoy typically applied science or social science backgrounds, many architects, designers and even some urban planners have a strong artistic bent. There is a very long tradition of looking upon city design as an art.

Yet urban designers have more clearly articulated the ideological and moral dimensions of their work. In the city the moral imperative is more immediate and more people are at risk. Hence Berleant's (1992: 81) statement that

> it is through creating an urban environment that is a dynamic synthesis of the practical and the aesthetic, where need and awareness are equally fulfilled, that function is both most complete and most humane, and where enlightened aesthetic judgement can become a social instrument toward a moral goal.

Unlike most nonurban landscape appraisal methods, urban planning procedures can readily be divided into qualitative and quantitative. What they initially have in common is an overwhelming devotion to the visual, despite decades of pleas for a multisensory urban aesthetic (Burchard 1957, Lynch 1976, Porteous 1982, Berleant 1992; see also Chapters 2.3 and 5.1).

Qualitative urban design

The aesthetic goal of traditional urban design methods is to produce an urban fabric which will delight those who experience it. Hence the initial step is to understand how urban places are experienced. The bulk of the work discussed here is normative, derived from designers themselves without formal public input.

Classic approaches

Rather than take a chronological approach to methodological critique, which worked fairly well for nonurban appraisals, it is salutary to look at a series of classic works on urban aesthetic form and to tease out their commonalities (Williams 1954, Burchard 1957, Rasmussen 1959, Lynch 1960, 1976, 1981, Cullen 1961, Cunliffe 1969, Whittick 1970, Lozano 1974, Smith 1974, 1977, Jakle 1987). Across this thirty-year span there is an amazing degree of agreement, partly because the later writers build on the work of the earlier. Consensus occurs particularly in terms of: (1) the basic structural elements of townscape; and (2) the fundamental aesthetic qualities of townscape. Both these components relate quite strongly to the typologies developed by experimentalists (Chapter 3.2) and landscape appraisers (Chapter 5.1).

The basic structural elements of city form are, in terms of Lynch's (1960) terminology (with that of others in parentheses):

* districts (sites, spaces, places)
* edges (lines of life, lines of contrast)
* paths (streets, exposures, circulation facilities)
* nodes (enclosures, open spaces, foci, centres)
* landmarks (focal points, monuments, individually significant architecture).

Although Lynch was most concerned with orientation in the city, the aesthetic importance of these visual elements comes out in all the texts reviewed. Similarly, all are concerned with views, variously subdivided into floorscapes, groundscapes, vistas, skylines, and panoramas. The water and vegetation content of townscape is also emphasized (see Chapter 3.3). Further, almost all texts emphasize that cities are usually experienced while in motion; they variously refer to the importance of movement, motion experience, dynamic vision, sequencing, rhythm, or serial vision.

When moving through the assemblage of form components that constitutes a city, the observer's degree of aesthetic delight depends upon the artistic coordination of a series of basic aesthetic qualities. Again, the reviewed texts tend to agree that these include, at the more physical, measurable end of the spectrum, formal, spatial and surficial qualities such as scale, proportion, mass, void, line, tone, colour, texture, and visual richness. At the more experiential end, the major qualities are the now-familiar (Chapters 2.1, 3.2, 5.1) unity, variety, complexity, coherence, dominance, ambiguity, enclosure/exposure, mystery, and surprise.

To take only one of the more recent of these works as an example, Jakle's (1987) list of elements that make up 'character in landscape' include:

- scale, which involves both hierarchy (city, street, building, architectural elements) and the relationship between buildings and human form; 'landscapes properly scaled impart feelings of [personal] adequacy' (p. 78)
- detail, which includes variations in architectural elements, vegetation, street furniture and the like, which enable the observer to savour differences and similarities in the rhythms that they create; 'much of the visual success of Baltimore's renovated waterfront is the result of light-fixtures, benches, planters, signs, kiosks, pieces of sculpturing and bollards all placed in careful juxtaposition' (p. 81)
- style, which relates to the appropriateness of detail and implies a characteristic, readily-recognized arrangement of things. Romantic style, for example, introduces 'spontaneity, tension, contraposition, juxtaposition, and randomness', while classical landscapes 'are distinguished by clearly intelligible compositions' (pp. 87–9, see also Chapter 2.2)
- rhythm, which enhances a sense of appropriateness, and includes serial rhythm ('a straightforward repeating of objects or object relationships in a metered cadence') and binary rhythm ('dialectic between opposites in the sequence of repetition: height/depth, constriction/openness, or darkness/light'). Or rhythm can be seen as regular, variable, or irregular (pp. 93–4). Whereas Romantic townscapes involve irregular rhythms, classical ones demand regular or variable rhythm, the latter involving an increase or decrease of cadence in a regulated manner
- face, which resides in the façades of buildings, and involves issues of viewpoint, frontality, profile, texture, and grain. 'Coarsely textured façades along a street tend to arrest the eye ... a complex face enhances the public space by inviting pause' (p. 102)
- light, which includes atmospheric changes and light conditions, light angle, reflection, and especially colour (hue, brightness and saturation). 'Colour is a dictator of mood ... a colour mood can invite action or inaction' (p. 105)
- change, an issue dealt with in great detail in Lynch's often-overlooked *What Time is this Place?* (1972). Change includes the great natural seasonal and diurnal cycles as well as the daily, weekly and seasonal activities of humans and the sense of time that comes from both historic urban fabric and ongoing construction projects.

An urban designer must therefore take into consideration 'awareness of scale, sense of appropriateness, search for face, concern with light, and concern for change [to create] a macroview variously composed of panoramas, vistas, focal points, enclosures, enclaves, and points of pause' (Jakle 1987: 115).

It is the task of the design team of architects, landscape architects, designers, engineers, and planners to produce a design which fuses all these elements and qualities into a coherent whole and, avoiding the extremes of monotony or chaos, to weave buildings and spaces together 'in such a way that drama is released ... for the city is a dramatic event in the environment

... a tremendous human undertaking' (Cullen 1961: 8). 'Bring people together', continues Cullen (p. 7), 'and they create a collective surplus of enjoyment; bring buildings together and they can give a visual pleasure which none can give separately.'

Clearly, townscaping is the art of relationship. It is already clear that simple standardized check-list approaches, such as are successfully used in nonurban landscape appraisal, are hopelessly inadequate as aids to the design of cities. In urban design, unfortunately for those who wish to standardize the world, there is still a premium on knowledge, experience, imagination, skill, and creativity (Burchard 1957).

Design as composition

The chief method of urban aesthetic design, then, is the art of composition. In other words, in aesthetic terms cities are treated as visual compositions analogous to works of art. This is an old tradition (Bacon 1976) and involves the notion of 'sight-thinking' rather than 'word-thinking'. The essence of a successfully composed view is a sense of unity coupled with legibility (Jakle 1987). Legibility is the Lynchean notion of an 'imageable' city composed of an integrated set of understandable districts, edges, paths, nodes, and landmarks. Unity is a proper balance of diverse elements which together appear to be all of a piece. The aspects of composition which promote unity include enframement (which sets a view apart from other views), a sense of entry and exit (which enables the eye to enter and leave the view comfortably), a focal centre (which holds the eye), and balance (which provides a sense of stability). Although most urban landscapes are not deliberately composed as works of art (though certain sections of city may be), a sense of beauty is likely to emerge when, inadvertently or otherwise, satisfying compositions appear, and are recognized and delighted in.

For a more dynamic view of townscape, one must return to the classic *View from the Road* study (Appleyard *et al.* 1964), which aimed not merely to improve the safety of American roads, but also to richly supply them with curves, signs, vegetation, landmarks, art objects and other road furniture which would, via the operation of change over both space and time, enhance drivers' opportunities for aesthetic pleasure. Cullen's classic *Townscape* (1961, originally 1949) appeared even earlier, and remains one of the more insightful manuals for both understanding and designing urban form. Cullen is wholly concerned with vision, and especially with the serial vision of the pedestrian, for whom 'the scenery of towns is often revealed in a series of jerks or revelations' (p. 9). Pedestrians are typically exposed to a series of views gained by turning corners, entering tunnels, alleys, or doorways, and emerging from closure into courtyards and squares. Aesthetically satisfactory sequences are created by linking existing views (hereness) with emerging ones (thereness) via the manipulation of colour, texture, scale, style, uniqueness and the like.

One of the best examples of this sequencing of views is provided by Thomas Sharpe, town-planner for Oxford after World War II. Sharpe (quoted in Tuan 1993: 112) describes a pedestrian's experience of the city centre:

> As he approaches the Bodleian from the top of Catte Street, there is nothing to see but its noble cube. Advancing, he sees first the rotunda, then the spire of St. Mary's, then the dome of the Radcliffe [Camera] coming into view. As this vast circular bulk separates from the bulk of the Bodleian, the tower of St. Mary's also emerges. Despite the fact that each of the three buildings is in its own way as sophisticated a piece of architecture as there is, the experience is elemental, beyond the power of words or photographs to describe. Cube, cylinder, and cone are suddenly juxtaposed . . . with a result that is . . . sensational.

Here is a case of 'sight-thinking' that can only be verified by a visit to Oxford. It is fully in accord with Cullen's townscape approach.

The chief value of the *Townscape* volume is its casebook, which explains Cullen's concepts in a series of diagrams, photographs and drawings with minimal text. His revival of the classical 'outdoor room' concept, for example, which is manifested in squares, plazas, and gardens, is based on the concept of 'hereness' which itself involves elements of closure, enclosure, multiple enclosure, looking in/looking out, closed vista, incident and mystery, and can be contrasted with 'thereness', which has its own attributes (Figures 5.10 to 5.13). One of the best aspects of Cullen's approach is this detailed visual vocabulary. Using such a handbook, the visitor or researcher can describe an existing townscape. With these concepts in mind, aesthetically sensitive new designs can be produced.

Cullen is amusingly sardonic when describing the usual standardized townscape designs, which he calls 'prairie planning'. If buildings were letters of the alphabet, he notes, then in most modern designs 'they are used not to make coherent words but to utter the monstrous desolate cries of AAAAA! or OOOOO!' (p. 133). Relph (1987) succinctly explains that such monotonous cityscapes have come about through the processes of standardized planning procedures, 'clean-sweep planning', 'planning by numbers', corporatization, and commodification. Modernist cityscapes he finds to be dominated by rational order and inflexibility, hardness and opacity, and discontinuous serial vision, where views and perspectives are marred by sharp disjunctions (Fig. 5.14). Sewell's (1993) recent attack on modernist planning in Toronto wholeheartedly agrees. Post-modernist townscapes are somewhat better in their reconnection with the local setting via the development of 'quaintspace' (Fig. 5.15), a sequence 'of enclosures, winding passageways, little courtyards, canopies over sidewalks, easy transitions and continuity of appearance between exterior and interior spaces' (Relph 1987: 253). Quaintification, although often inauthentic, does permit what Cullen calls 'the possession of territory', so that people linger and even perform, and the city once more becomes theatre.

Figure 5.10 Streetscape with terminal punctuation: York, England

Figure 5.11 Entrypoint with mystery (and surprise): Holy Trinity, Goodramgate, York

Figure 5.12 Enclosure , with distant goal: Sommières, France

Figure 5.13 From hot openness, via cool archway, to closedness:
Ioánnina, Greece

Figure 5.14 Uncoordinated modernist urban landscape: Vancouver, BC

The established design traditions represented by Lynch (1960) in the United States and Cullen (1961) in Britain respectively promote coarse-grained (automobile) and fine-grained (pedestrian) planning. Their mutual aim is to design cities aesthetically from the point of view of a moving person, so that 'the whole city becomes a plastic experience, a journey through pressures and vacuums, a sequence of exposures and enclosures, of contrast and relief' (Cullen 1961: 10). The explicit aim of the townscaper, then, is to 'manipulate the elements of the town so that an impact on the emotions is achieved' (p. 9). Lynch extends this goal to the scale of the urban region, providing a well-illustrated handbook for sensory quality planning aptly entitled *Managing the Sense of a Region* (1976). This work on regional planning has been extended by Hough (1990).

Although there appears to be no substitute for such manuals, coupled with exemplary case experience, traditionalist urban designers have been both assisted and challenged by the urban scale-model movement. Although several city models exist, notably in Sweden, Israel, Holland and New Zealand (Francis 1987), the best-known was developed at the University of California, Berkeley, in the early 1970s. The concept involves building a scale model of the city, district, or building complex in question. Then, to avoid the 'helicopter view problem', an upside-down periscope device is used to enter the model and permit the observer to experience views from any point. Coupled with a camera and a moving gantry, the periscope can take trips

Figure 5.15 Post-modern quaintspace, Victoria, BC

through the model and record these in a variety of media. Thus proposals for urban (re)development can be built into the model and their aesthetic effects observed. Proponents see this tool as valuable for all types of designers, as well as for providing information to the public and politicians and for enhancing environmental education (Mellander 1974).

The model of San Francisco built at the Environmental Simulation Laboratory in Berkeley (Fig. 5.16) was used, for example, to develop a set of sun-access guidelines designed to promote the penetration of sunlight into the city's streets, parks, and other public open spaces (Bosselman, Flores and O'Hare 1983). Based on such guidelines, cities can generate sun-access zoning plans which limit building heights and promote light- and warmth-enhancing building orientations and shapes.

Related work has been concerned with the growing 'New Yorkization' of San Francisco, whereby what was a unique, low-rise, open, pastel-coloured Mediterranean-style city in 1960 had become a dark, forbidding high-rise identitown (Chicago? Singapore?) by the end of the decade (Fig. 5.17). This proliferation of Modernist cubes (Ford 1994) had untold effects, including problems of safety, stress, social interaction, and maintenance. The aesthetic effects are chiefly a matter of severe visual disruption via the height, bulk, colour, shape, scale, and detail of the new buildings, and especially their frequent blocking of pleasant pre-existing views of harbour, water, vegetation, and hills. The Berkeley simulator proved useful here in simulating high-rise alternatives, one of the major goals being to prevent homogenized tower proliferation from further reducing the uniqueness and specific character of

Figure 5.16 The model San Francisco in the Berkeley Environmental Simulation Laboratory

the San Francisco townscape (Appleyard and Fishman 1977). Related social surveys demonstrated that high-rise developments reduce residents' aesthetic satisfactions and, as aesthetics appears to be a major influence on people's evaluation of an area, high-rises also promote indifference and even hostility to the urban scene (Dornbusch and Gelb 1977a). Similarly, high-rise development has a negative effect on the enjoyment of parks (Dornbusch and Gelb 1977b) and the liveability of neighbourhoods (Gelb 1977).

STOP THEM FROM BURYING OUR CITY UNDER A SKYLINE OF TOMBSTONES

Both the above pictures are of downtown San Francisco. Same spot. Same time of day. Same weather conditions. The top one was twelve years ago. The bottom one, last year. San Francisco was once light, hilly, pastel, open. Inviting. In only twelve years it has taken on the forbidding look of every other American city. Forty more skyscrapers are due in the next five years. They are as great a disaster for the city economically as they are esthetically. Ask a New York taxpayer.

What can you do to stop it?

Contact SAN FRANCISCO OPPOSITION, 520 Third Street (second floor) or telephone 397-9220.

Figure 5.17 The 'New-Yorkization' of San Francisco 1960–70

In sum, 'satisfaction with the visual and living environment in San Francisco will decrease under high-rise forms of development unless they more specifically address people's aesthetic objectives' (Dornbusch and Gelb 1977a: 111). It is significant that a number of North American cities, including San Francisco, California, and Victoria, British Columbia, have introduced curbs on tall building development. Models like the Berkeley simulator, coupled with the growing use of colour video systems and computer graphics (Francis 1987) with their overtones of public disclosure and participation (see Chapter 5.1) and even citizen protest (Chapter 4.2) introduce us to the possibility of involving public preferences in the urban and regional aesthetic design process.

Planning for public preferences

Visual surroundings which appear so chaotic as to defy comprehension, and those which contain such incessant repetition as to be monotonous, are to be found throughout the urban scene. Aesthetic satisfaction can hardly be experienced under such conditions, and the resulting 'slow grinding toll on our nerves suggests the urgency ... of doing something about these conditions' (Williams 1954: 95). Williams called for an aesthetic vision which eschewed concern for city elements (parks, civic centres, and the like) in favour of a comprehensive approach to the aesthetic design of the city as a whole. His suggested method was first, a qualitative three-dimensional analysis of form, texture and colour, and second, an analysis of the way cities are perceived. Williams's sense of urgency was a response to Cullen's earliest townscape work, published in 1949, but he advanced beyond Cullen by calling for public participation. Nevertheless, very little in the way of public preference work was performed before the late 1960s.

The 1960s were indeed a time of crude economic determinism. Because of the heavy reliance on the easily-measured economic benefits of development, which too often ignored less quantifiable environmental costs, the liberal economist J.K. Galbraith (1966: 57) was moved to declare that 'we must explicitly assert the claims of beauty against those of economics'. Initially, attempts were made to bring economics and aesthetics together in methods of public preference elicitation involving evaluator models, bidding games and the like (Pendse and Wyckoff 1974, Randall, Ives and Eastman 1974). These attempts were, however, rather crude compared with the sophisticated phenomenological design games used by modern geographers (Jungst and Meder 1986), which will, perhaps, eventually result in a useful scenic grammar.

Psychology was a more common route towards public preference. One of the first goals was to (re)discover the 'quality components' of urban form that are likely to produce delight or pleasingness in the observer. Not surprisingly, the findings generally match those from the nonurban scene (Chapters 3.2, 5.1), notably the importance of ambiguity, complexity, variety, and unity (Rapoport and Hawkes 1970). Based on Gibson's (1966) notion that we perceive via borders and orderly changes in a state, Rapoport and Hawkes suggested that aesthetically pleasing city form required a hierarchical series of scales. Changes in scale, surprise, pattern and texture dominate the discourse of the architectural sculptor Cunliffe (1969), who claims that 'everything should be *very* instead of *somewhat*' (p. 163) and that 'we want to invent a magical world that we can walk around in' (p. 167). Research quickly developed and influential books such as Venturi's post-modern treatise *Complexity and Contradiction in Architecture* (1966) promoted more general interest in public perceptions of the urban scene.

By the mid-1970s, however, more sober appraisals pointed to the continuing, perhaps permanent, gap between research and professional practice (a hallmark of decades of discussion in such bodies as The Environmental Design Research Association) and the notorious inability of architectural criticism to 'generate common criteria and acceptable methodologies of analysis to evaluate built forms and spaces' (Lozano 1974: 353). Aesthetic judgement of cities by professionals, says Lozano, is still highly subjective, its vocabulary basically consisting of a superficial polarity of negative (monotonous, regimented, chaotic, confused) and positive (balanced, clear, interesting, dramatic) adjectives. Lozano therefore called for more sociobehavioural and psychological research (which still rely heavily on polar adjectives in semantic differential scales!), with strong concentration on the two issues of greatest importance, orientation/unification (the need for simple structuration) and variety/rhythm (the need for visual complexity).

Methods

The chief problem with such studies becomes one of measurement. In terms of successful measurement capabilities, Whitbread (1977) analysed three kinds of urban environmental quality studies: the behavioural, which looks at how people react to change (move away, soundproof the bedroom, raise or lower house prices); the experimental, where community members participate in controlled gaming simulations; and social surveys, where respondents express opinions in response to questions, often on bipolar semantic differential scales (good/bad, beautiful/ugly, etc.).

Problems with the first method are legion, including its indirectness and inferential character. Using house-price models as an example, Whitbread shows that unambiguous measures for explanatory variables can rarely be found. He investigates a variety of simulator models, including the Berkeley model, a British simulated living-room with manipulated external traffic-noise levels, and a 'noise machine'. The chief problem here is the difficulty of ensuring that respondents are reacting to the aspects of the simulation salient to the research question. Direct public preference elicitation via semantic differential scales and adjective check-lists, in contrast, produces more direct and comprehensive information. Whitbread suggests that the most valuable approach would be a combination of questionnaire and simulator methods. He concludes, however, that 'subtleties of theory regarding the assessment of social gains and losses as a result of public programmes which alter urban environments are almost totally academic in the face of the practical measurement problems involved' (p. 109).

Those who disagree include a large number of environmental psychologists who have made highly sophisticated quantitative analyses of public perceptions of neighbourhood visual appearance, urban roadsides, streets, parks and beaches, and downtown cores (Peterson 1967, Peterson and Neumann 1969, Preiser 1973, Winkel 1973, Byrom 1974, Anderson and Schroeder 1983, Im 1984, Beer 1991). Much of this research was reported in Chapter 3. We are no longer surprised to find that people prefer visually coherent cities with moderate levels of environmental complexity, dislike mess and chaos, and rate water, vegetation, and views very highly. Of note, however, is a significant body of research on the importance of gardens to urban people (Ulrich 1976, Francis and Hester 1988, R. Kaplan and S. Kaplan 1989), to which might be added information on the general 'greening' of cities (Andresen 1978) and the creation of urban forests (Miller 1979). I remember my delight in 1984 on finding that much of the downtown area of Managua, Nicaragua, flattened by an earthquake and not rebuilt because the relief funds were stolen by the dictator Somoza, had been turned into a forest park.

The best approach, then, would be a combination of simulations and questionnaires, both of which are fitted by Bagley (1975) into a 'general systems theory' of urban aesthetic assessment. This demonstrates how a carefully-chosen set of procedures can be used to evaluate alternative designs on the basis of how well they satisfy both project and community objectives.

One of the more interesting recent developments is the successful use of nonurban landscape-preference techniques for the elicitation of urban aesthetic preferences. For example, Im (1984) used the Scenic Beauty Estimation method developed by Daniel and Boster (1976), while Anderson and Schroeder (1983) used similar 'wildland scenic assessment methods' to ascertain public preferences for city scenes. In aggregate terms, these studies found strong respondent agreement about preferred and unpreferred scenes, suggesting the value of such methods to city planners. Anderson and Schroeder conclude that 'the urban environment is as conducive to aesthetic study as the natural environment' (p. 236) although much more allowance must be made (qualitatively, I interject) for the importance of historic, cultural, and symbolic values. Im asserts that landscape preference methods are valid and reliable, and castigates urban planners for their continued reliance on qualitative methods, an outcome of their continued belief that the measurement of urban visual quality is impossible. He suggests (p. 255) that 'the problem of visual quality measurement can be replaced by ... visual preference measurement'.

Recent reviews of environmental design research (Nasar 1988, Groat and Després 1991) confirm the value of such approaches, while demanding improved techniques and emphasizing the different worlds of discourse inhabited by planners and the public. Confirming the continued existence of this gap, Francis (1987) notes that although visual quality remains one of the highest priorities for designers and bureaucrats when creating city spaces, it is frequently given a much lower rating by users. Riley (1987) therefore calls for more study of 'the vernacular landscape', where the social and experiential are of greater importance than the merely aesthetic. Riley demands that both researchers and designers acknowledge the 'range and complexity of human affective response to landscape, its variability both among and within cultures and individuals, and the effect of time and exposure on those responses' (p. 146). In this connection some useful work on how Americans conceptualize their residential environments has been begun by Hummon (1990).

Applications

As usual, 'more research is needed'. The chief problem from the planning point of view, however, is how to translate researchers' findings into useful planning procedures. In a review of Nasar's useful edited collection *Environmental Aesthetics: Theory, Research and Application* (1988), Whitehead (1989) concludes that environmental aesthetics research fails to meet its praiseworthy goal of helping professionals to design settings which better fit public preferences. Environmental psychologists consistently claim that their work will be useful to planners (Canter 1977, Nasar 1988); equally consistently, planners tell them that their work usually comes too late, that is, at the analytical stage rather than the conjectural, goal-forming stage of design (Hillier, Musgrave and O'Sullivan 1972, Whitehead 1989).

That this is too sweeping a condemnation is seen from a review of three recent books. Altman and Zube (1990) provide designers with some useful background information on the way cities work both behaviourally and aesthetically. More immediately relevant is Francis and Cooper Marcus's *People Places* (1990) which, eschewing the old site planning where guidelines were presented as a set of rules, addresses the earliest, conceptual stages of design by inviting designers, researchers, and community groups to think about the specific problems of both a site and its users before developing a site-design programme. Beer (1990, 1991) demonstrates that a new approach to city design is increasingly demanded by politicians, the public, and professionals. She believes that the key issue lies in bridging the gap between the rather inaccessible information produced by researchers and the practical needs of designers. But although she cites a few 1980s research studies which produced information which fed directly into the planning process, she shows that the recent governmental (re)awakening of interest in the quality of urban life (European Community 1990) demands that first, researchers must understand how planners and designers operate, and second, that the latter must be trained to understand where to find the research information they need to help them make decisions.

Thus the main issue, for urban and nonurban aesthetic planning alike, is not primarily technical or methodological, but the more intractable one of improving communication between the public (source of information and the body to be served), the researchers (who tap and analyse the data), and the planners (who use this information to guide responsible decision-making). Were these lines of communication significantly improved, the urban site-planning process (Lynch and Hack 1984, Beer 1990) would be able to cease using 'best guesses' about probable user satisfactions, an approach which has led to massive planning disasters (Hall 1982) and the ultimate demolition of several major projects (Cooper Marcus and Sarkissian 1986). There is a huge body of evidence, from pre-project simulations to post-occupancy evaluations, which is of use to planners; the failure of urban designers to use this material is appalling. The disdainful godlike architect is apparently still alive and well.

Practical results

The practical results of several decades of interest in urban aesthetic planning translate into: a wide variety of better-designed city environments; the institution of design controls for new developments; and the creation of conservation strategies for existing urban fabric. It would be tedious to enumerate specific cases of good aesthetic planning; volumes of case-studies are available. Middleton (1987), for example, provides a compendium of interesting American, Australian, British and French approaches. Bad aesthetic planning, however, continues apace. Chevalier's *The Assassination of Paris* (1993), for example, is a passionate attack on the destruction of the face of Paris by a secretive, faceless, and totally confident technocracy. Kutcher's *The New Jerusalem* (1972) is interesting for its analysis of the opposing views of 'deductivist' and 'inductivist' planners in terms of their attitudes to history, values, and aesthetics. Whereas the inductivists value a 'civilized and humanly-scaled continuation of our existing environment', eschewing 'isolated works of brilliance' (p. 166), deductivists demand the fullest possible freedom of artistic expression for the individual architect. As Kutcher notes, 'a deductive Jerusalem would replace visual determinants with the inevitable operation of economic and technological necessity, leaving the city's form to chance ... unfortunately, if past experience is any guide, "open-ended form" and "on-going process" in concrete terms means visual chaos and environmental desolation ...' (p. 84).

It is to prevent these excesses on the part of arrogant architects and their overweening clients that cities across the Western world have developed design controls. As Burchard (1957: 117) remarked: 'A great urban aesthetic arises not from a cluster of architectural *chefs d'œuvre* but from a sensitivity on the part of each successive builder to the amenities already there.' Burke's *Townscapes* (1976) for example, provides numerous well-illustrated case-studies of new buildings which, implanted insensitively in existing townscapes, have ruined the visual quality of whole streets and districts.

Design controls, then, are public policy attempts to ensure that existing townscapes are respected when changes are made, and that new developments conform to at least a minimal level of aesthetic quality. Qualitative judgements suggest that new buildings should be 'well-mannered' in relation to the qualities of the existing urban fabric. Burke (1976: 175) shows how in London the 'unmannerly insertion of a bulky block of flats (1930s) into an Adam terrace (1775)' not only shaved off part of an existing building's pediment and destroyed several pilasters, but, more importantly, 'rudely interrupted the flow of streetscape'. Burke stresses the importance of preserving the visual qualities of everything from doorknobs and small shop fronts through district streetscapes to urban silhouettes. For instance, the distinctive skylines of Paris and London have been brutally destroyed since the 1950s by a proliferation of modernist and post-modernist blocks. Despite extensive work on skyline planning (Attoe 1981), the unique historic cities of the world are becoming more and more the visual equivalents of Kansas City.

In an attempt to provide a public-preference foundation for design controls, Harrington and Preiser (1980) performed a study which concluded that the general public: (1) is quite capable of judgement and agreement on aesthetic issues; (2) can readily identify and reasonably judge visual aesthetic subcategories such as scale, height, massing, set-back, landscaping and the like; and (3) can provide useful input into the formulation of visual aesthetic guidelines. Nevertheless, existing guidelines have generally been created without much public input and vary tremendously in scope from city to city. Houston and Amarillo, Texas, have no aesthetic controls, whereas the most detailed and comprehensive controls are to be found in such cities as Brighton (UK), Canberra (Australia), Calgary (Canada) and Santa Barbara and Palo Alto (California).

Preiser and Rohane (1988) then analysed the architectural control mechanisms of sixty-four cities in the United States, Britain, Canada, Australia, and New Zealand. Regulation proved to be highest in British-influenced countries and most lax in the United States, as might have been expected (Chapter 2.4). Whereas in the latter country concern was generally restricted to set-backs, signage, land-use compatibility, height, and fencing, the other countries had city ordinances covering such visual elements as sunlight and daylight access, exterior lighting, modulation or rhythm of solids, voids, and fenestration, design subordination to valued landmarks, landscaping, and public art. The frequency of use of aesthetic control elements by the sixty-four cities sampled appears in Table 5.2. Perhaps the most significant result of this review is the discovery that although urban aesthetic controls exist only at the local political level, there is a strong commonality of approach across the five English-speaking nations (with similarities also in Scandinavia).

Preiser and Rohane suggest several plausible reasons for this agreement, and conclude by asking if it might be feasible to 'create an international

Table 5.2 Frequency of urban aesthetic control issues

Item	Regulated aesthetic issue	No. of the 64 surveyed communities citing issue
1	Signage restrictions	56
2	Height restrictions	54
3	General compatibility of building	52
4	Off-street parking and loading zones	50
5	Conservation of existing vegetation	48
6	Specific landscape elements required	45
7	Percentage of open space required	44
8	Volume and massing review	44
9	Materials and texture review	44
10	Colour and finish review	36
11	Style and character review	34
12	Maximum and minimum size of buildings or lot	34
13	Visual air-pollution regulations	30
14	Underground-utility requirements	30
15	Percentage of site to be landscaped	29
16	Silhouette or profile review	27
17	Façade-articulation review	26
18	Public art required or encouraged	11

Source: Preiser and Rohane (1988: 246), courtesy of Cambridge University Press

master set of aesthetic variables' (p. 430). They note, however, that the control of aesthetic urban quality is only possible 'if a certain willingness to be regulated exists in the population' (p. 430). And the existence of such a willingness, I contend, depends in part upon improved environmental education (see below, Chapter 5.3).

Before discussing that topic, however, it remains to demonstrate the relative success of heritage conservation in cities (see also Chapter 4.4). Whereas the impetus for conserving nature developed in the 1960s, it was not until the 1970s that the preservation of the urban past became a widespread concern in Europe and North America. In a distinctive turn obvious today to any city resident or visitor, the 1950s and 1960s push for unthinking development of new buildings and freeways gave way in the 1970s, after much activist protest (Chapter 4), to a preserving and restoring movement which has revitalized threatened 'old towns' not only in such visually pleasing tourist cities such as San Francisco, Vancouver, and Victoria, British Columbia, but also in areas once considered obsolete and valueless, such as the docklands of London, Liverpool, and Hull. Lowenthal (1979) concludes a useful review of 1970s activity with the remark that 'things worth saving need not necessarily be beautiful or historic as long as they are familiar and well-loved' (p. 555), which anticipates the discussion of 'valued environments' below (Chapter 5.4). Similarly, Ford (1974) demonstrates that people like San Francisco's Jackson Square chiefly for its variety, texture, depth, scale and visual contrast, and declares that 'the old criteria for preservation [glorying in the past] are no longer appropriate as spatial, functional and visual factors take precedence over historicity' (Ford 1979: 211).

A whole range of urban conservation planning guides, manuals and reviews appeared in the 1970s (Ziegler 1971, Astles 1972, Lottman 1976, Goodey 1977, Dobby 1978, Artibise and Stelter 1979, Newcomb 1979, Jakle and Travis 1980), all aiming to prevent the spread of what Relph (1976) calls 'placelessness'. Some of the best of these guides are written by geographers, who have a far wider concept of historical preservation than do architects, archaeologists, or art historians. Those who doubt the value of conservation, or question its cost, are referred to Newcomb (1979) who, ignoring the basic question of why we should preserve 'heritage' (for which see Lowenthal and Binney 1981), takes a market-oriented approach based on practical experience in Britain, Denmark and the United States. As Simms (1982: 293) remarks: 'There is a lot of money to be made by the planned use of historical features.'

Many local studies are available. An example of interest to me is Kluckner's thoughtful, witty, and provoking *Paving Paradise* (1991), an assessment of heritage gains and losses in British Columbia. Kluckner details the ongoing destruction of 'leafy, soft-edged, rather "English" neighbourhoods' in Vancouver by the brash arrogance of *nouveau-riche* 'monster houses' which uproot trees and expunge traditional landscaping to create an urban nowhere (Fig. 5.18). It is good to discover environmentalists, heritage conservationists, and neighbourhood-quality movements coming together in the early 1990s to force Vancouver city council to enact a 'neighbourhood character' by-law to prevent such excesses.

In contrast, Victoria city council's acceptance of Cadillac-Fairview Corporation's demand to build a downtown mall, known as the Eaton Centre, in the mid-1980s, signalled a failure to preserve an attractive city-centre building assemblage. Victoria is an attractive city with many pleasant streetscapes, few high-rises (there is a height ordinance), a pedestrian-friendly environment, and much original quaintspace. Until the 1980s it was one of the few large Canadian cities without a huge downtown mall. Not only has the resulting Eaton Centre destroyed many existing businesses, including a smaller adjacent mall, but it has also produced a bizarre example of heritage-style veneerism (Fig. 5.19). Despite the fact that their chosen site contained unique historic buildings in a variety of styles, Cadillac-Fairview refused to listen to a vociferous public protest, and bulldozed the city council into rejecting all alternatives except demolition and replacement.

Unlike façadism, which implies the retention of at least the outer walls of graceful old buildings (Figure 5.20), veneerism involves constructing a completely new replacement edifice in a pseudo-traditional style. The result in Victoria, which I try to avoid, is an extraordinary 'time-warp' experience. At first glance, the pedestrian does not notice the obvious newness of the gimcrack detail of the Eaton Centre's exterior, where fake structural members such as fibreglass columns and corners vie with fake Venetian

Figure 5.18 A 'monster house', Victoria, BC

Figure 5.19 Veneerism: the Eaton Centre, Victoria, BC

Figure 5.20 Façadism: heritage skin retained, Victoria, BC

blinds on fake upper-storey windows. Kluckner (ibid. p. 40–1) eloquently describes the experience of this building:

> Downtown Victoria's architectural continuity is so reassuring that, although Eaton Centre is much shinier than the carefully restored historic buildings on the adjoining streets, it is possible to walk through the downtown immersed in the ambience of old shops and streets and then, with scarcely any awareness of what is about to happen, to walk through the doors into Eaton's. Suddenly, you are transported from historic Victoria to Yorkdale in Toronto, or Pacific Centre in Vancouver, or a new suburban Eaton's Centre anywhere else. The colours shift so suddenly from muted bricks outside to the tropical turquoise of 1990s mall-land that it is quite disorienting. Few architectural composites in North America create such an intense transition.

The only positive aesthetic result of this post-modernistic fakery was the sale of part of a street to Cadillac-Fairview, which generated enough cash, ironically, to set up a Downtown Heritage Building Incentive Program! To put all this into international context, the reader is referred to Middleton's (1987) well-illustrated volume of conservation case-studies.

It remains to note that heritage conservation is no longer restricted to cities. Archaeologists, foresters, and architects, ironically for geographers who once claimed the region as their unifying concept, are developing schools of regional architecture, regional archaeology and the like (Tainter and Hamre 1988). Urban and regional planning have long been associated. Hence there is a growing literature on the conservation and rearrangement of historic landscapes in both Britain (Goodey 1977, Brandon and Millman 1978) and North America (Wright and Hilts 1993). It is notable that, in Britain at least, recent years have seen a number of popular television programmes on the conservation of rural and urban landscapes, culminating in the designation of 1992 as 'Landscape Year'. While Goodey's (1977) proposal that communities should institute their own 'community heritage audits' has not received widespread response, national inventories of conservable resources are ongoing in many countries.

Summing up

As a researcher, I conclude that the prevailing emphasis among researchers on methodological exactitude is self-defeating. Brilliant research methods and 'useful' research findings are useless unless there is improved two-way communication between researchers and planners. Zube (1991) calls on social scientists to be more aggressive in getting involved in local, regional, national and international environmental programmes. And no planning is really worthwhile unless the dialogue is extended to include public participation, with public preference studies as the minimal acceptable input. Further, separate streams of urban and nonurban landscape appraisal are counter-productive, hampering theoretical advances, methodological refinement, and implementation.

Even if all the above problems were solved, aesthetic planning will be of little value unless public involvement increases, a goal which could be fostered by environmental education (Chapter 5.3). And, ultimately, environmental aesthetics cannot be improved if powerful countervailing interests get in the way (Chapter 5.4).

5.3 ENVIRONMENTAL EDUCATION

The preceding discussion of urban and landscape planning deals essentially with the short-term. Even public policy in the form of national legislation (Chapter 4.4) may be considered short-term in the sense that we are still unable to ensure that planners give aesthetic 'intangibles' sufficient consideration, while the monitoring of business corporations for infractions of environmental guidelines wastes considerable manpower and has resulted in numerous lawsuits. The chief problem, therefore, becomes one of implementation and enforcement.

Enforcement would hardly be necessary if human beings learned to eschew short-term goals and become tuned in to the eternal. If people could be induced to regard both nature and human places as sacred sites, there would be no problem of nest-fouling. Unfortunately, the Western world in the Reagan–Thatcher years of the 1980s seemed in no mood to accept the small-is-beautiful point of view (Schumacher 1973), the god-in-nature arguments of eco-theologians (geologians?) (Barbour 1972, 1990, Cupitt 1984, Frye 1991), or the Gaia hypothesis (Lovelock 1979) as viable models for living one's life. The revival of environmentalism in the early 1990s, however, gives rise to hope for an ecologically sound and aesthetically pleasing future.

Nevertheless, it is realistic to assume that eternalist attitudes to environment will grow only slowly. The immediate goal is to replace short-term attitudes to environment, which generally see it as a commodity, with long-term attitudes which look to a sustainable rather than a maximal level of 'progress' and development.

The route to the long-term, of course, lies through the vale of education. It is clear that activism, through its claim on media attention since the 1960s, has raised the consciousness of the general public about environmental aesthetic issues. Sustained interest in environmental aesthetics and ecology, however, is likely only if a long-term programme of environmental education is instituted at all levels of education, especially the kindergarten and elementary levels. The argument here is, in the words of St. Paul, 'to train up a child in the way that he should go', with sensitivity training in environmental issues and perhaps a dash of what Schumacher (1973) calls 'Buddhist economics', which emphasizes the importance of 'right livelihood' and reverence for life.

If consciousness-raising could be accomplished on a grand scale, goes the argument, there would be far fewer problems of implementation or enforcement, for citizens would 'naturally' care about their environments and become ethically incapable of harming them. In terms of Figure 1.3, the goal is to develop in the average citizen an attachment to the earth as the home of humankind, a traditional geographical concept that took wing, once again, with the first views of the earth as a whole, from space, in the late 1960s. It is no coincidence that the American landing on the moon was closely followed in 1970 by the first Earth Day. A balanced combination of attachment, spirituality, ethics, and aesthetics would, it is believed, ensure and sustain the human presence on the earth.

To this end, three basic modes of environmental education have been promoted. It is to be understood that these are interdependent holistic approaches which complement, rather than replace, the more traditional scientific approaches of geography, geology, biology, and environmental studies. They are: environmental interpretation; environmental autobiography; and landscape sensitivity training.

Environmental interpretation

Environmental interpretation is the deliberate explanation of an environment, usually to visitors (Sharpe 1976, Goodey 1977). The movement grew rapidly in the 1970s, and is chiefly American in origin, mainly concerned with recreational rather than everyday environments, and to be found in rural and wilderness as well as urban areas. Tools for interpretation include displays, leaflets, wayside markers, descriptive labels, trail guides, and films and talks given by park rangers at park headquarters. The concept was later extended through the development of self-guided town trails, which rapidly became popular in Britain.

There is an enormous literature on environmental interpretation (Halprin 1969, 1970, Aldridge 1975, Pennyfather 1975, Percival 1979, Herbert 1989, Uzzell 1989, Wilson 1991), and the last decade has witnessed an explosion of interpretative provision by the managers of 'heritage' attractions. It is felt that people need to know more about what they are looking at, especially the context, history, and wider significance of particular historic or ecological sites (Prentice 1991). Yet we know little about the designer's interpretative objectives, which may be to entertain as much as to inform. Indeed, some critics (Bennet 1988) argue that the 'heritage' presented, in Britain at least, is value-laden, partial, and unreal, and may involve a hidden agenda. In practice, some visitors have found the approach involves too much cut-and-dried instruction, and a few find labels attached to woodland trees to be offensively redolent of botanic gardens.

School group visits seem especially ineffective, for children regard these as opportunities for release and fun; I have watched Israeli children happily frolicking in that most sober of Holocaust museums, Yad Vashem. Very little effort has yet been made, however, to discover what is learned from interpretation and how it is learned. Pioneer work by Prentice (1991) involved taking parties of undergraduates to Welsh castles which were heavily interpreted by wall-mounted display boards, or 'books on walls', a major thrust of Cadw, the Welsh state agency responsible for monuments. Student recall of on-site interpretation via multiple choice tests (which even a chimpanzee could do well on) was appalling. Prentice concludes that: the objectives of on-site interpretations need to be clarified; we need to know how different audiences respond to interpretation; and 'the determinants of gaining the attention of visitors ... need to be researched' (p. 307). In other words, we as yet have little information about the effectiveness of environmental interpretation.

Wilson's provocative book, *The Culture of Nature* (1991) traces the history of environmental interpretation in North America, but has little good to say about it until the recent trend towards a more proactive, ecologically-sound, conservationist type of interpretation. Earlier efforts in Canadian national parks, for example, involved constructing profiles of visitor types and dividing parks into 'interpretation management units' to accommodate them. Only recently have interpreters become anxious to engage visitors in crucial issues such as environmental politics. Wilson outlines the difficulties involved in effective interpretation and suggests that biospheric systems and their politics are 'not easy to explain to people on holidays in a way that will immediately speak to their hearts' (p. 60).

Goodey (1978b) tackles this problem by suggesting that the interpretation movement has concentrated too exclusively on the didactic instructing of a supposedly ignorant public. He asserts that every site needs an interpretative plan as a working document, but that the chief purpose of interpretation should not be to instruct, but first, to provoke, and second, to relate what is being described to something within the personality or experience of the individual visitor.

This is clearly a tall order, especially at a site which may receive thousands of visitors annually. Yet interpretative labels could certainly be more imaginative than a dry recitation of scientific nomenclature; a label on a willow tree, for example, should indeed tell us that this a species of *Salix*, but might then go on to tell us that the leaves of this tree are the original source of aspirin. In this connection the Costa Rican National Parks Service deserves commendation for its signage. In one volcano and forest park I was charmed by two provocative signs:

- Indians lived here for 20,000 years and never left any litter.
- This tree was here before Columbus arrived in America. How old are you?

Other trees bore poems of an arboreal nature.

In recent years, the major thrust in interpretation has moved away from direct instruction, which has come to be regarded as a possible hindrance to the visitor's need to explore and find things out in a more personal way. Berleant (1992), for example, castigates art museums as lifeless collections of disparate objects which fail to promote a lively encounter with art. He makes interesting suggestions for the redesign of both buildings and exhibition space so that the museum of fine art is transformed 'from collections of objects into opportunities for aesthetic engagement' (p. 123). Creativity and imagination, coupled with some knowledge of how visitors actually use museums, are therefore necessary.

These criticisms have provoked salutary changes in both indoor and outdoor interpretation in recent years. Clearly, the major task of interpretation is now seen to be the effective guidance of both residents and visitors toward opportunities for personal discovery and experience.

Environmental autobiography

Whereas environmental interpretation seeks to open people up to environments they may not normally experience in everyday life, environmental autobiography reintroduces us to places we know so well that we have forgotten how to articulate what we know.

Elsewhere (Porteous 1989) I have demonstrated that autobiography is little written or researched by academics who, their brains clouded by the myth of objectivity, regard the genre as subjective and self-indulgent. Yet autobiography allows outsiders to get in, and is especially valuable in trying to understand the motives of those who transform our public environments. And, according to the philosopher Dilthey (Thompson 1978: 45) 'autobiography is the highest and most constructive form in which the understanding of life confronts us'. Hence my otherwise thoroughly academic book *Planned to Death* (1989) incorporates relevant fragments of my geoautobiography.

I call 'geoautobiography' any autobiography which deals extensively with the relationships between person and environment. Many childhood-oriented autobiographies have a strong sense of place. As Cobb (1975) remarks in one of his accounts of 'the geography of my childhood', 'to write about a place is also to write about the people in it and about yourself'.

In a world which everywhere tends to negate the individual (Ellul 1981) we nevertheless recognize that some landscapes still have identifiable 'authors' (Samuels 1979). Samuels asks about 'the *who* behind the image or facts of landscape', and calls for an investigation of those who do the authoring. This is the biography of landscape. Environmental autobiography takes a different tack by asking those who create environments to investigate themselves.

An American movement in origin, environmental autobiography is closely associated with schools of architecture and landscape design, and particularly with the University of California at Berkeley (Ladd 1977, Cooper Marcus 1978, 1979, Helphand 1979, Hester 1979). The reasoning behind the generation of environmental autobiographies on the part of students in design classes is to provide an opportunity for future professional designers to regain contact with their own experiences. Too often, graduate students are taught to discount their personal experience, while at the same time they remain unaware of how strongly their past may influence the character of their designs.

With an extended studio presentation in mind, students are encouraged to tap their memories and recapture or even revisit environments that were significant in their past lives. In many cases, the process is salutary and self-revealing in a productive way. Very rich material often emerges. Instructors and students alike may be struck by 'the depth of our own past environmental experience which we hide or ignore when observing or making decisions about the present' (Goodey 1982).

According to Helphand (1979: 12):

The process of presenting one's environmental history acts as a transition between 'intuitive' and 'conceptual' understanding. As one might expect, often the intuitive experience is a validation of theoretical concepts.... Instead of being taught a set of ideas, conceptual understanding is discovered latent in individual experience and the shared experience of others. The discovery process both validates and helps imprint ideas.

In other words, self-exploration liberates the environmental imagination, understanding results from doing, and self-knowledge improves future receptiveness to environmental experience and points out both personal blocks and personal biases in design solutions. Given these advantages, environmental autobiography might be a salutory exercise in self-revelation for anyone. As Bachelard suggests in *The Poetics of Space* (1964: 23) 'We should undertake a topo-analysis of all the space that has invited us to come out of ourselves ... each one of us should make a surveyor's map of his lost fields and meadows.'

It is clear that individual creativity is strongly influenced by context, especially that of childhood (Buttimer 1993). Bourassa (1991) demonstrates how personal aesthetic strategies overcome biological and cultural constraints. Jungst and Meder (1986: 209) declare that the life-history of researchers and planners

is constitutive for their work ... such persons are especially prone to become enslaved in their subject, to lose their detachment from it, thus tending to produce artifacts conditioned by their own life histories. Both researchers and planners must be capable of keeping their life histories under control in the research and planning process.

Awareness of one's idiosyncracies would therefore seem to be a *sine qua non* for anyone involved in the transformation of environments. Alas, many blunder blindly on, burbling about 'progress'.

Environmental sensitivity training

In contrast to the two approaches discussed above, environmental sensitivity training is chiefly a British development. Its aim is to develop a sensory awareness of place on the part of schoolchildren, and to enable them to develop a 'feel' for the built environment. It has chiefly been an urban movement, and its attention to the positive aspects of cities is salutary in an urbanized nation that tends to reject the city in favour of the countryside in attitudes if not in residential location.

Based on environmental interpretation concepts, the notion of the 'town trail' has expanded into the concept of 'the multi-sensory exploratory trail or walk' (Goodey 1977). But because of the artificiality of providing interpretative instructions such as 'stop here and smell a flower', the interpretation has been transformed into such modes as the exploratory game. In this mode a pack of cards is used, comprising five sense cards, four movement cards (stretch, walk, run, move), four directional cards (left, right, turn around, forward) and other cards demanding action (ask, tell, draw, write, imagine, take from the environment, add to the environment, and the like). The participant child shuffles, deals himself a card, follows the instruction, and then deals another card after an appropriate distance interval.

Similarly, the Town and Country Planning Association began in the 1970s (Goodey 1978a) to create manuals on how to set up sensory walks for children, 'town trails without facts'. The sequence involves:

- Thinking time (provoking quotations);
- Loosening up (loosening the body and its senses; blindfolds, for example, liberate the non-visual senses);
- Feeling places (imagining places where you might feel happy or sad, or where you might run, or hide);
- The sensory walk itself, during which children are alerted to underfoot sidewalk texture, pedestrian choreography, smells, sounds, weather, clothes, trees, colours, and art, and experience running, dawdling, asking the way, and the 'chance dance' of intersections.

The goal is to increase awareness and provide a foundation for personal growth as well as for descriptive, analytical and experimental activities.

Children may be encouraged to take an interest in aesthetic environmental issues in many other ways, of course. Delinquent children have been rehabilitated by such simple means as providing them with video equipment on which to record their neighbourhood or city centre. Children respond positively to the 'streetometer' concept, which permits them to evaluate the world built by grown-ups in terms of litter, wirescape, floorscape, advertisements, parking, vegetation, building condition, road safety, air freshness, noise, and a host of other variables. These developments are outlined in great detail in Wheeler and Waites (1976). Ward and Fyson's *Streetwork: The exploring school* (1973) takes an explicitly radical position in its effort to encourage children to realise that their environments are not immutably fixed, but are man-made, often badly, and can and must be changed. A 'Learning Through Landscape' programme is now in place in British schools (Lucas 1993). In Canada, Stanley King's townscape design work with children is outlined in the National Film Board's *Chairs for Lovers*, while Moore (1980) has involved children in the evaluation of open spaces.

There is no need to deny adults the environmental enjoyment that we expect in children. In this connection the designer Lawrence Halprin (1969, 1970) developed 'town scores', which score a walk as if it were music, using observation and questioning techniques very similar to those used by British children on sensory walks. When asked in 1970 to develop design concepts for the town of Everett, Washington, Halprin chose to involve Everett's citizens in 'a purposeful set of environmental explorations – on foot, by car, and by helicopter – during which "untrained" observers could both sense and evaluate their everyday environment from a number of different viewpoints'. Halprin's 'environmental scores' are designed to promote urban design which structures activities and experience but allows considerable room for improvisation. The chief aim is to establish 'a common language of awareness', in much the same way, in another but related field, that Christopher Alexander *et al.* has attempted to develop *A Pattern Language* (1977) for architecture.

Individual adults, of course, may undergo sensitivity training through a personal exercise of the will, without instruction from experts. Indeed, self-training, with minimal guidance, may be the best option for adults. In my undergraduate Environmental Aesthetics course, I encourage students to write me a description of a landscape; those who are still untainted by the social science craze for objectivity often produce the most interesting essays. They go out into the field armed with a ten-page manual entitled 'One Hundred and One Questions for Reading a Landscape' (provenance unknown; it was introduced to me, in fact, by an undergraduate). Each question comes with a drawing, a provocative quotation, and a paragraph of suggestions. The questions begin with the senses, range through a variety of physical characteristics reminiscent of Cullen's *The Concise Townscape*, explore history, politics, and behaviour, ask about feelings, empathy, and 'what the place says', and culminate with critique, including a critique of the observer's own process. The feedback from this exercise is generally good. Some essays are so evocative that I am moved to make a special effort to visit or revisit the places described.

Less structured guidelines for better seeing are also useful, and are most readily found in modern interpretations of John Ruskin and other pioneers. As the subtlety of the argument defies brief synopsis, I refer the reader to, *inter alia*, Berger's *Ways of Seeing* (1972) and *On Looking* (1980), Aiken's 'Toward landscape sensibility' (1976), and Relph's 'To see with the soul of the eye' (1979), as well as to the discussion above in Chapter 2 and below in Chapter 5.4.

And yet

Aesthetic activists have been successful in bringing the issue of environmental aesthetics squarely before the public. Since the 1960s a rash of legislation in North America, Europe, and elsewhere has attempted to include aesthetic factors in modern environmental quality standards. Yet cultural inertia, the gap between environmental words and environmental deeds (O'Riordan 1976), and the consequent continued existence of environmental ugliness, suggest that the time has come to give more effort to the longer-term strategies of environmental education, with the hope that more eternalist views of landscape might thereby emerge.

The emergence of certain North American wilderness landscapes as 'ceremonial landscapes' (Erickson 1977) or 'sacred space' (Graber 1976) augurs well, for attitudes towards such landscapes appear to be basically religious. For certain groups, at least, ecology has attained religious status, with its sacred books (Thoreau, John Muir, Aldo Leopold), ritual expressions of devotion in the shape of wilderness pilgrimages, and the development of codes of personal conduct which are known broadly as 'green'.

Although such attitudes still raise wry smiles among many, it is likely that only by the widespread development of an eternalist ecological spirituality will the earth be retained as an environment in which life is worth living. And aesthetics remains a major component of this attitude of reverence toward the earth.

In this connection, the activist Heath (1974) reviews a vast amount of research which demonstrates that aesthetic appearance is a vital factor in people's reactions to both indoor and outdoor environments. Yet, he says, architects and planners have until recently tended to avoid the word 'aesthetics', preferring more *macho* pseudo-scientific terms such as 'environmental quality'. But as aesthetics seems to be a basic need, we will have to become brave enough to say 'aesthetics' in front of children. Significantly, Heath's forceful article is entitled: 'Should We Tell the Children about Aesthetics, or Should We Let Them Find Out in the Street?'

5.4 AESTHETIC CRITIQUE

Continuing the theme of eternalism, it is reassuring to note that modern theologians are developing a theology of environment in which the aesthetic appreciation of landscape has a prominent place (Frye 1983, 1991). While Paul Tillich sees modern life in Western culture as permeated with a profound sense of meaninglessness, Martin Buber expounds the difference between an I–It attitude toward nature, which is exploitative and uncaring, and an I–Thou attitude which involves a feeling of human stewardship toward the land. Such a feeling of stewardship is both explicit and implicit in Judaeo-Christian culture (Kay 1989), contrary to those who have frequently expressed the view that this prevailing Western religious base is fundamentally hostile to nature (White 1967). In accord with the former view, the philosopher Martin Heidegger calls for an attitude of environmental humility (Relph 1981).

Psychiatrists such as Harold Searles (1961) have also begun to see the non-human environment as 'one of the most basically important ingredients of human psychological existence'. Childhood environments, in particular, have a powerful influence on personality development (Coe 1984, Porteous 1989, 1990). Searles asserts that North American culture fosters a detached, unhealthy view of environment; it is above all a mobile culture, detached from roots and place and always ready to discard the old in favour of the new. A recent critique by von Maltzahn (1994) shows how thoroughly we have divorced ourselves from the natural world. In this connection, Hough (1990) suggests that the classification of scenic landscapes and so-called objective ratings of the visual superiority of one landscape over another merely highlight our society's disassociation from the land. Yet a basic human need, as Maslow (1954) also noted, is for relatedness, and this includes relatedness to the earth in the shape of gardening, hunting, walking, and other pursuits. Such a positive relationship with the earth, notes Searles, is likely to improve relationships with other people. But to improve our earth relationships, we must first learn to see.

Learning to see

Among social scientists, geographers have been most prominent in promoting landscape appreciation. Meinig (1971) defines environmental appreciation as the act of being keenly sensible of or sensitive to one's surroundings. Aiken (1976) has similar views. Meinig envisions landscape appreciation regaining its eighteenth-century prominence, so that it becomes as important as the appreciation of art and music, and like them, may be enjoyed at a wide range of levels. Seeing must become a humane art rather than a merely analytical science, and as such must involve art and literature. For Meinig (p. 4) the 'skilful novelist often seems to come closest ... in capturing the full flavour of the environment. His sensitivity to a scene, to the seasons, to the special qualities of life in a particular locality are often vividly evocative.'

Literary geographers (Pocock 1981, Porteous 1985, 1990) would agree, and one thinks particularly of Jane Austen's *Sense and Sensibility*. Appleton (1980) sees the study of landscape as a fertile meeting ground for both the arts and the sciences. And the applied arts and sciences, he might have added.

Above all, however, 'environmental appreciation is an art, it is holistic, it is particularistic, peripatetic, qualitative, sensual, and, ultimately, idiosyncratic and deeply emotional' (Meinig 1971: 11). In other words, landscape appreciation is life-enriching and very personal, and as such will have difficulty in being taken seriously in schools or universities in our scientistic and instrumentalist age.

Most critics calling for a citizenry more appreciative of landscape stress the importance of learning how to see. Lewis (1979: 31) believes that one should 'teach oneself how to see, and that is something Americans have not done'. Meinig (1971: 5), by contrast, urges 'greater training in how to look at an area, how to open the eyes and become keenly sensible of one's environment'. (And, as we have learned in soundscape and smellscape studies, to open the ears and nose also.) Similarly, Tuan (1994) argues that 'the eye needs to be trained so that it can discern beauty when it exists'. There is no indication, however, of how this eye-opening training might be accomplished.

Relph (1979) provides some clues. *Pace* Ruskin, he asks us to 'see with the soul of the eye', but stresses that the responsibility lies with the individual; 'we have to teach ourselves to see landscape aesthetically' (p. 31). The key to such seeing is 'understanding and insight based on clear observation and attention to the aesthetic properties of scenes'. Specifically, 'we must always ask ourselves why this scene looks as it does, what functions it serves and whether it serves them well, what values it expresses, and how it relates to neighbouring landscapes' (p. 31). More specifically still, the exercise '101 questions for reading a landscape', as noted above (Chapter 5.3), provides a very detailed means of interrogating a landscape for its aesthetic values and meanings.

Insights into landscape can reciprocally result in insights into the observer's own nature. The end result of many such individual experiences could well be a general public realization of the I–Thou relationship with nature. But even if this proves impossible, says Rees (1975: 46), 'we must learn to tolerate and, if possible, to like or love nature, if not in spirit, then aesthetically'.

To achieve this state he asks us first to attempt to empathize with the early nineteenth-century Romantic movement whereby nature was revered by savants who were as much scientists as artists. Yet the truth these Romantics sought from nature was not scientific but moral and aesthetic. They appear to have had an emotional engagement with the landscape which the twentieth century has lost.

The main need is to learn to see well, and for this environmental and aesthetic education is required. Visual education has three conditions. First, a sharpening of our perceptions, an enlarging of our range of vision beyond the merely functional. Second, a heightening of our sensibilities which must currently be quite dull because otherwise we would not tolerate so much environmental ugliness. Third, there must be a realization that the vision of the artist has as much validity as that of the scientist.

I would add other prerequisites. It would be valuable, I feel, in increasing sensibility among the young, for a more widespread teaching of both the history of landscape aesthetics and a more rigorous training of children in art and music. Children should also be encouraged toward more adequate expression of their landscape sensibilities through art and music making. Central European children have long been exposed to curricula rich in artistic and musical appreciation and technique. This is certainly not the case in Britain and North America. And not only children; professionals and scientists might well be encouraged to express their appreciation of environment as musicians, painters, novelists, poets, or cooks. Both the geographers Watson (1983) and Meinig (1982) and the humanistic psychologist Maslow (1954) have called for alternative forms of self-expression by academics. Specifically, Tuan (1961: 32) opined that 'Geographers ... might well take time off from their duties and join – at least now and then – the artists and poets in portraying the splendor of earth'. I have attempted to put these visions into practice through the publishing of geopoetry (Porteous 1984, 1988).

Humanist critique

Having learned to see, we must then learn to critique. Meinig (1971: 23) demands:

> A much larger body of citizens whose eyes have been opened to see their surroundings with far greater appreciation, and who thereby are not only enriched by what is good, but appalled by that which is bad, and refuse to countenance the continued despoliation of their surroundings.

In refusing to countenance this despoliation, they clearly must learn to critique it in a coherent and incisive manner. Much of Chapter 4.1 was devoted to humanist critique of environment. Here it suffices to note that recent thinking in environmental aesthetics is wholly in favour of active critique.

Bourassa (1991: 132), for example, follows Carlson (1977, 1990) in rejecting too unthinking a reliance on public preference surveys, calling instead for a corps of landscape critics, creative environmental professionals who have 'both an in-depth understanding of landscape processes and a sensibility that allows [them] to see in unconventional ways'. Similarly, Sancar (1989) has argued that architectural criticism should be extended to cover the whole landscape, a form of 'critical regionalism' that has long been the province of humanistic geographers (Lewis 1979). Berleant (1992) begins his chapter on 'environmental criticism' by suggesting that such an activity hardly exists at present, at least in a sense comparable to art or literary criticism. He feels that architectural criticism cannot be extended to landscape because it is too concerned with individual buildings or objects without place-relationships, it is dominated by formal, technical, and external considerations, and its 'star system' of overboosted egos has little relevance to landscapes created, often quietly, by many hands. Finally, architectural criticism is wholly visual and concerned with external appearance. In contrast, to appreciate landscapes we enter them.

Instead, Berleant calls for a wide variety of approaches to environmental criticism, including 'textual, semiotic, formal, moral, mythical, psychological, and social and political interpretative frames' (p. 136). The objectives of such criticism are manifold. First, as with art criticism, the major goal is to develop and enhance appreciation. Second, criticism can draw attention to the aesthetic dimension of environment, 'so that it assumes an equal place with other, more commonly recognized environmental values – economic, conservation, historical, moral' (p. 144). And third, of course, critique can support positive developments and denounce bad ones, thus promoting efforts to design environments in ways which recognize the practical importance of aesthetic values.

One of the more positive and less elitist suggestions emerging from humanistic environmental-aesthetic critique is the concept of valued landscapes. In relation to both evaluative and public preference landscape assessment techniques, several researchers have noted that an individual's aesthetic appraisal of a landscape may be as much bound up with factors such as familiarity, comfortableness, affection, and attachment as with judgements of beauty or attractiveness (Penning-Rowsell 1981a, b, Sansot 1983). These authors also suggest that appraisal practitioners operate on too broad a scale, for people are chiefly concerned with local aesthetic issues, not the regional or national ones which exercise the minds of planners. Further, in their emphasis on a few concepts such as variety and uniqueness, current planning approaches simply carry on the century-old tradition of selecting spectacular landscapes, to the detriment of ordinary pastoral scenes which, though loved by their inhabitants, go unprotected and are thus subject to heavy development pressures (see Chapters 2, 4). Similarly, urban residents have a rich relationship with their city, unlike the fleeting gaze of the tourist or the abstract visions of architects, designers, and engineers (Rimbert 1973).

The concept of valued environments, places that ordinary people appreciate and want to protect, was developed by geographers in the late 1970s. Lowenthal (1978a, b) is quick to point out that aesthetic preference is only one factor in attachment to landscape. Although Tuan (1974) identified certain 'environments of persistent appeal' (tropical forests, seashore, valley, and island), Lowenthal's historical and spatial account demonstrates the sheer variety of landscapes to which people may become attached. Hence 'to ascertain what locales are specially favoured and why calls for an understanding of a country's history and institutions' (p. 62), an understanding much more likely to be humanistic than scientific.

The variety of valued landscapes has been explored in a volume edited by Gold and Burgess (1982). Several contributors explain how individuals and groups come to value particular places, and how far they are willing to protect these from external assault. Similarly, Relph (1993) suggests that planning should recognize that places are made through the involvement and commitment of the people who live and work in them. Planners and designers should therefore serve not as experts, but as 'environmental midwives' in the place-making process. Hester (1993) shows how this can occur in practice, explaining a North Carolina community-led planning process designed to enhance economic development while preserving locally valued places.

Of course, the 'valued environment' concept is clouded by Norberg-Schulz's (1981) suggestion that every place has its *genius loci*, experienced by its residents as 'sense of place'. Hence, protecting special landscapes, which is presumably the goal of this planning process, means protecting all landscapes. And Moffit (1975) has troubling thoughts on the meaning of 'value'. Nevertheless, this concern with what is valued by ordinary folk, as opposed to experts, whether critics or not, echoes George Orwell's firm belief that 'there must be a place in the modern world for things that have no power associated with them, things that are not meant to advance someone's cause, or to make someone's fortune, or to assert someone's will over someone else' (Shelden 1991: 436). Which brings us to radical critique.

Radical critique

A more radical critique of contemporary landscapes is provided by Marxist and other socialist aestheticians. Debord (1967), for example, sees modern society becoming increasingly a society of spectators, where almost every experience is no better than second-hand. Postman's (1982) critiques of television converge here. Capitalism has unified space, resulting in the extensive banalization of places, the generation of placelessness (Relph 1976), and the creation of blandscapes (Porteous 1990). Tourism, 'human movement considered as consumption ... is fundamentally nothing more than the leisure of going to see what has become banal' (Debord 1967: 168). Debord views economic history as developing largely around the opposition between city and countryside. Moreover, modern 'urbanism destroys cities and re-establishes a pseudo-countryside [including low-density suburbs] which lacks the natural relations of the old countryside' (p. 177). There is thus a total loss of aesthetic sense and of natural rhythms. The modern suburban environments 'of the technological pseudo-peasantry' are aesthetic nothings and historical zeros: 'On this spot nothing will ever happen, and nothing ever has' (p. 177).

Thus, a radical approach to nonurban aesthetic landscape appraisal might well dismiss the whole exercise as fundamentally misdirected. By the year 2000, about half the world's population will be urban. A more balanced landscape assessment practice would devote greater time to townscape

assessment. Scarce research resources would be redirected from the evaluation of remote scenic areas which most citizens will never see, or only rarely.
If, instead of ignoring the urban scene, we could improve the citizen's
aesthetic satisfaction with it, there might result not only a happier citizenry
but also a lowering of pressure on nonurban areas through a decline in out-
of-city discretionary travel. And if recent findings on the anxiety-producing
effects of city life are validated (Chapter 3.3), the case for a reorientation of
the research effort towards the everyday urban environment is even greater.

More radical still would be an approach which suggests that urban
aesthetics pales into insignificance beside issues of class (Walker 1981), the
increasing economic and spatial scale of property development (Eichler and
Kaplan 1967), the influence of real-estate interests (Weiss 1987), the circulation of money (Harvey 1985), and the engorgement of the financial sector
(Walker 1988). Hence Jones (1991: 242) argues: 'In landscape planning, the
cultural landscape is a moral landscape. It cannot be divorced from
dominating ideologies in planning and political circles.' He therefore urges
a continual critique of the exercise of power. Tuan (1993: 226) in a milder
tone, confirms that 'the good and the beautiful, the rural and the aesthetic,
are inextricably intertwined', but he has no revolutionary plan to foster the
creation of good and beautiful landscapes or to prevent the ongoing
production of the bad and the ugly.

We live in a world of 'contested realities' (Symanski 1994) and 'contested
global visions' (Cosgrove 1994). Despite the collapse of so-called Communism, revolutionary thinking is still valid. 'Proletarian revolution', writes
Debord (1967: 2) is nothing less than 'the critique of human geography'. The
critic Raymond Williams (1973) agrees, noting that landscape is not neutral
and cannot, as happens so often with the humanist approach, be understood
apart from its social, economic and political context (Figures 2.14, 5.21).
Socialist critique asserts that landscape appearance is primarily a matter of
property ownership. Landscapes have biographies (Samuels 1979) and these
biographies are inextricably intertwined with the economic, social, and
political desires of powerful individuals and elite groups, whether
eighteenth-century noblemen or modern multiglomerate corporations (for
an example of the latter, see Porteous 1989).

Landscapes reflect power (Figures 2.19, 2.28, 5.21). Landscape, viewed
aesthetically, is an expression of personal or corporate wealth and the
consequent power to dictate patterns and forms. Landscape implies a
generous amount of leisure time on the part of the observer; it implies
observation, separation, and a degree of detachment. For this reason, a
working district, to a working person, is hardly ever a landscape (Williams
1973).

Both Williams, in his analyses of literature, and Barrell (1972, 1980) in his
critique of landscape paintings, make the point that the elegant townscapes
and serene parklands of eighteenth-century noblemen, still much admired
today, were founded on the toil and suffering of a large stratum of urban and
rural labourers. To create unimpeded prospects whereby landowners could

Figure 5.21 A landscape of power: Thomas Gainsborough's *Mr and Mrs Andrews*, 1748 (reproduced by courtesy of the Trustees, The National Gallery, London)

actualize and reinforce their sense of power and control, roads were stopped up or realigned, commons enclosed, churches taken into private parks, and whole villages destroyed. As the novelist and art critic John Berger has so tellingly shown in *Ways of Seeing* (1972), at the heart of a good many contrived landscapes, paintings, and pastoral poems lies the worm of injustice.

Watson (commentary in Moggridge 1986: 113) is more direct: 'certain activities actually destroy landscape. Most of these are concerned with money.' Watson goes on to speak about the wickedness of landscape destroyers (p. 113):

> ... the evil selfishness of generations of landowning gentry ... totally without responsibility, consideration, or care. Society has of course learned from them, and their rottenness has spread: to ruthless estate-builders, factory-farmers, and litter-louts. All these anti-landscape people are merely copying the example of the English nobility and landed gentry.

Cultural trickle-down with a vengeance!

In the United States, Tunnard and Pushkarev (1963) have similarly demonstrated how American landscapes are achieved not by public consensus but by the wealth and power of small elite groups and their tame professionals. Kurtz (1971, 1973) asserts that the resultant beauties of landscape must not be regarded as mere contextless objects, but should be related back to the motives of those who caused the landscapes to be made. In Britain, Crosby (1967, 1973) sees landscape as an expression of the growing authority of those who hold the largest stakes, namely governments and large corporations. And it is worth noting that both power and the landscapes it produces are still very much the prerogatives of males. It does not bode well for environmental causes of any kind that we live in a gendered economy with a gendered bureaucracy; for while women tend to view power as a means to promote change, men use power as a means of having influence over both people and things.

These mostly male-oriented vested interests are frequently hostile to public beauty (Sancar 1985), and have become much more so since the radical right, symbolized by Thatcher in Britain and Reagan in the United States, came to power in the 1980s and began to change the world to suit big business. Those who can declare, like Thatcher, that 'society does not exist', are hardly likely to be sympathetic to social demands for attractive and healthy environments, especially if these demands counter the market, demand public intervention, or lower business profits. It is possible, for example, for the radical right to claim that the 1989 *Exxon Valdez* oil-spill disaster was really a benefit to Alaska, because the environmental clean-up it engendered provided jobs and increased GDP. Significantly, Ridout (1988) attributes the refusal of Wisconsin to adopt 'Scenic Beauty Rules' as being due to 'market failure', opposition by industrialists, and the political climate of Reaganism. The considerable business and political opposition to environmental movements of any kind is epitomized by the speech of United States

Presidential candidate Patrick Buchanan on the first night of the 1992 Republican convention. 'Environmentalism was explicitly linked with radical feminism, atheism, homosexuality, and all the other evils of the Democratic party. It was opposed to Freedom and the Judeo-Christian Ethic and the Right of Working Men to Have Jobs' (Katcher and Wilkins 1993: 190).

Unfortunately, many believe such trash, thus considerably slowing the efforts of activists to encourage the aesthetic amelioration of environments (Chapter 4). Planners and academics, too, are implicated.

Hence Tunnard's 'tame professionals' have also come in for considerable criticism. Planners may be hampered by corruption (Wachs 1985) or process limitations. All too often they operate as scientists, and for J.G. Ballard (in *Love and Napalm*, quoted in Lasch 1984: 139) 'science is the ultimate pornography, analytic activity whose main aim is to isolate objects or events from their contexts in time and space', an impossibility when dealing with environments. Or they operate as technologists, which is equally unsuitable for environmental or aesthetic understanding. As Postman (1993: 158) states:

> In Technopoly precise knowledge is preferred to truthful knowledge [and] in any case Technopoly wishes to solve, once and for all, the dilemma of subjectivity. In a culture in which the machine, with its impersonal and endlessly repeatable operations, is a controlling metaphor and considered to be the instrument of progress, subjectivity becomes profoundly unacceptable. Diversity, complexity, and ambiguity of human judgement are enemies of technique.

Even the social science-based work on public preferences falls into the scientistic trap, with practical consequences. For 'social researchers ... see themselves, and can be seen, as scientists, researchers without biases or values, unburdened by mere opinion. In this way, social policies can be claimed to rest on objectively determined facts' (Postman 1993: 158). And both scientists and planners are elites, divorced from public perceptions and arrogantly self-perpetuating. Hamill (1989) suggests that the excellent reputations of landscape practitioners such as Leopold stifle criticism of and refinement of their methods. He further suggests that prevailing establishment beliefs that objective knowledge is possible and that quantitative knowledge is superior to any other leads to the making of, and persistence of, fundamental errors. Moreover, innovative approaches are not encouraged unless they conform to the basic philosophical stance of objectivity and quantification. Peer review also comes in for criticism, for

> in the literature of landscape analysis, the ideological, technical or other affiliations of individuals may usually be identified with reasonable accuracy in the items included in footnotes and reference list, as well as in the items excluded. The ... non-recognition of critics is as important in protecting individual and group interests as the positive support of adherents.
>
> (p. 203)

Hamill concludes that 'many of the institutional practices of universities [and other institutions, one might add] seem to be intended mainly to identify and reward orthodoxy' (p. 204).

Landscapes, then, are symbols of vested interests (Figure 5.21). They are, at bottom, ideological expressions, and assist in the task of imposing the world-view of a dominant class upon the remainder of society. Aesthetic appreciation of landscape, therefore, is bound up with a deep, subtle political process of societal identification with elite, male-dominated interests. Society as a whole is 'bought off' via consumer commodification and a rampant materialism which, ironically, uses landscapes as a major feature in what has become a discrete genre of false and misleading environmental advertising (Dunham 1994). Clearly, in this view, the aesthetics of landscape is a surficial mask below which lies a reality of dubious class and gender ethics.

Conclusion

Both Marxists and liberals, if not conservatives, have in the past subscribed to the vision of a leisured society which would hold creativity in high esteem and would endeavour to foster such creativity among all types of citizens. Such a consummation has not yet come about, despite the premature heralding of the so-called 'leisure society' in the 1950s. Indeed, our current worship of electronic technology in a society whose culture has become television (Postman 1982, Porteous 1990) suggests that this vision is never likely to reach fruition. Research demonstrates that even native North Americans, traditionally close to the land, are losing the kind of local awareness that television documentaries cannot supply; 'the personal, uninhibited, and spontaneous interaction with nature which solitude allows is seldom taking place today' (Nabhan and St. Antoine 1993: 240). McKibben (1989) and Katcher and Wilkins (1993) demonstrate that both television and science teaching rely on texts and pictures, all two-dimensional, while direct learning from the world around the child is increasingly devalued. Hence, 'after years of excellence in science, students enter medical ... school unable to distinguish reliably between veins and arteries' (Katcher and Wilkins 1993: 192). 'Learning to see' becomes a sick joke; for as people are technologized, their eyesight deteriorates.

With this 'extinction of experience' (Nabhan and St. Antoine 1993), and the gradual 'closing of the American mind' (Bloom 1987) associated with a rapid decline in cultural literacy (Hirsch 1987), together with increased teletechnology and the development of mass tourism (Figure 5.22) in an increasingly disposable world (Hackney 1995), it might seem all but inevitable that informed landscape appreciation will remain the prerogative of a well-educated elite.

Figure 5.22 Scenery of the future?

Bank of China building, Hong Kong

AFTERWORD

beyond aesthetics

We should do our utmost to encourage the Beautiful, for the
Useful encourages itself.

Goethe

No author feels wholly at ease when attempting to sum up a whole book in
a few sentences. But it is necessary, for this book does not merely attempt to
explain what has been done in Environmental Aesthetics, but also formulates
an argument, which I briefly recapitulate below.

Aesthetic appreciation of environment, chiefly visual and aural in humans,
may well have a universalist, survivalist, evolutionary base. Yet cultural and
social differences seem paramount when judgements of landscape beauty are
made, suggesting that aesthetic appreciation also involves meaning, and
therefore learning. Individual differences in environmental appreciation
appear to be merely variations on a sociocultural theme. Heuristically,
Environmental Aesthetics can usefully be understood as an interrelated
matrix of four major approaches or 'windows on the world', namely the
Humanist, Experimentalist, Activist, and Planning paradigms.

Humanists formulate concepts through observation and reflection. They
are responsible for the evolutionary theory of aesthetics and for its sociocultural counterarguments. Notions of the contribution of environmental
aesthetics to personal well-being are ancient humanist concepts. Despite many
conceptual and methodological problems, experimentalists have provided
much support for humanist ideas such as the 'nature tranquillity hypothesis'.
Their work also involves the creation of environmental assessment measures,
thus bridging the gap between humanists and planners.

Once minimal agreement on aesthetic issues appears in select groups,
idealistic activists of various stripes engage in action, from the illegal to the
legislation of public policy, to enforce aesthetic standards on those who seek
to alter public environments. More pragmatic, planners try to operationalize
experimental assessment methods and implement legislation incorporating
humanist concepts and activist concerns. Theirs is the most unenviable task,
for problems are inevitable as aesthetic concerns continue to be considered
hostile to the entrenched politicoeconomic interests of those who value

profit and having above pleasure and being.

At this point, as introductory overviews always say, we have merely scraped the surface of the emergent interdiscipline known as Environmental Aesthetics. Although this book has, I hope, promoted the growth of the infant interdiscipline, provided an overview of work already done, and formulated a typological framework for organizing and interconnecting that work, it is clear that this is only a beginning. In terms of research and practice, the vast array of topics to be included in the agenda of 'what needs to be done' includes: further investigation of the evolutionary theory of aesthetic appreciation; an attempt to rectify the lack of coherent theory in the interdiscipline; more consideration of the non-visual senses, with increased multisensory research; the mutual refining of concepts by qualitative humanists and quantitative experimentalists; improved methodology on the part of the latter, including a greater concern for working directly with planners in developing theoretically-sound, workable appraisal techniques; a rapprochement, long overdue, between landscape planners and urban designers; and better means of incorporating public preferences and values into the planning process in a world where public aesthetics is almost always subject to political and economic constraints, not to speak of widespread public indifference (Rimbert 1973).

In the long run, of course, we will have to humanize that world and banish that indifference. To achieve these goals, a major effort in education and improved communication is required.

Although I have demonstrated that there is considerable potential for the flow of ideas and information between the four major approaches to Environmental Aesthetics, and have suggested that good aesthetic planning depends on this intercommunication, it is also clear that there are many impediments to such interaction. Some of these are bound up with the hierarchical thinking of both academia and society at large, which still glorifies the Analytical Scientist (Experimentalist) mode of operation, thereby downgrading or even ignoring the seminal contributions of other approaches. Humanists, in turn, tend to be remote and uninvolved.

As apolitical academics working most often in an impersonal manner, both groups bear some responsibility for the 'applicability gap' which looms between empiricists and theoreticians on the one hand, and planners and the public on the other. And as we have seen, failure to communicate between planners and the public is endemic. The planners themselves face a yawning gap between landscapers and townscapers. The high value accorded to the nonurban environment by humanists, confirmed by experimentalists, and institutionalized by scenic resource managers confirms the anti-urban bias of our society. Even the theorists Appleton and Smith (Chapter 1) are distinguishable respectively by their rural and urban concerns.

Yet the discussion of planning in Chapter 5 above suggests that humanist, experimentalist and activist concerns are deeply implicated in planning thought and process, and that improved linkages between these approaches would be useful. Improved communication might result if academics and professionals were to become more aware of their biases, perhaps via exercises in environmental autobiography (Chapter 5.3). As Jung has remarked (Mitroff and Kilmann 1978), not only does each of his types have an opposite, but this opposite represents the undeveloped, unconscious, or suppressed side of that type. An Analytical Scientist, for example, may spend his life developing that one side of his personality, to the neglect of the opposite, Conceptual Humanist side, which if acknowledged might provide some balance in both life and work.

In a world of growing psychological interest and awareness, it does not seem unreasonable to explore the possibility of promoting increased self-awareness on the part of academics and practitioners. Both Analytical Scientists (Experimentalists) and Conceptual Theorists might be encouraged to take a more personal and political view of their work, which would perhaps lead them to consider context and applicability. Conceptual Humanists, with their schemes for general human betterment, might be encouraged to operate more locally and specifically, while Activists (Particular Humanists) would benefit from a more global outlook.

In terms of the several personality models outlined earlier (Chapter 3), we might be more optimistic about achieving an attractive, ecologically-sound earth if professionals became more generalist (Little), more conservationist (Phillips and Semple) and more active and holistic (Sancar).

Overall, a pessimist might conclude that the several approaches to Environmental Aesthetics are rife with problematic difference. An idealist might well recognize their essentially complementary nature and seek to break down the unnecessary barriers between them. Although the position of Environmental Aesthetics at the intersection of many approaches and interests is one of its major difficulties, it is at the same time one of its major strengths, for its goal is integration, interconnection, and a holistic appreciation of landscape.

Re-education, however, must not stop with academics and professionals. The way to a more aesthetically pleasing world with a far better distribution of aesthetic welfare (see Chapter 1) is improved public education in aesthetics, beginning at kindergarten (see Chapter 5.3). The general goal of aesthetic education would be to sharply expand the Activist constituency as a means of countering the efforts of capitalism to produce an ever more ugly, homogeneous and utilitarian world in which beauty must be either profitable or private.

The more specific goals of aesthetic education would include: recognizing the inextricable linkage between private and public lives and environments; abandoning the current *macho* approach to aesthetics of any kind, especially among males; and giving high priority to teaching children not merely utilitarian tools for making a living (or a killing!) but also aesthetic and ethical tools for living a full life. Like the ancient Greeks (Sancar 1985), we have to recognize that beauty is not independent of politics or ethics (see Chapters 1.1, 5.4).

Indeed, as Keats reminds us, ethics and aesthetics are irrevocably intertwined. And as Santayana so eloquently puts it in *The Sense of Beauty* (1896: 73):

> Not only are the various satisfactions which morals are meant to secure aesthetic in the last analysis, but when the conscience is formed, and right principles acquire an immediate authority, our attitude to these principles becomes aesthetic also. Honour, truthfulness and cleanliness are obvious examples. When the absence of these virtues causes an instinctive disgust, as it does in well-bred people, the reaction is essentially aesthetic.... It is *kalokagathia*, the aesthetic demand for the morally good, and perhaps the finest flower of human nature.

Similarly, Ruskin felt that one could respond to beauty not merely sensuously (*aesthesis*) but also with one's whole moral being (*theoria*) (Fuller 1988).

Should we fail to develop such an aesthetico-ethical consciousness, let us imagine the near future of Western urban-industrial nations, which could well involve vast ageing, retired populations watching television all day and dying of inanition because of poor life-skills and inadequate sociocultural support. This is a future to which many are condemned in capitalist societies organized by crude profiteers and their vulgar technological henchpeople, and where children are trained, rather than educated, in schools in which 'only a very tiny portion of formal education [is] devoted to the skills involved in the enjoyment of living' (Novak 1985: 247).

We are in this position because although we have made some headway in liberating ourselves from political and sexual repression, we have simultaneously volunteered for other repressions, notably consumer egoism, the tyranny of the expert, and the domination of the machine. The three come together in the arresting image of loving but tongue-tied parents laying their child down to sleep to the sound of a lullaby tape. In a world as highly educated as ours, there is surely no need to accept a pseudoculture which relies so heavily upon the opinions of experts or is dominated so completely by machinekind. A Victorian-style world of enthusiastic amateurs would be preferable. My leap of faith is that an educated citizenry, currently so entranced by the aesthetics of the personal, might well be able to carry this enthusiasm over into a more widespread concern for the aesthetics of the public environment. Even a widespread eternalist view of the world is not impossible, for I believe with E.M. Forster (1910: 275) that 'we are evolving in ways that Science cannot measure, and Theology dare not contemplate'.

In the long term one of these ways may be towards the development of an environmental religion, which would involve the recognition of non-economic environmental intangibles as major factors in the development and nurture of human identity, whether personal or group. These factors include attachment, aesthetics, ethics, and spirituality (Figure 1.3). To pursue this idea would take another book; the curious reader will find an outline in Porteous (1993). To elevate these factors to the prominence they deserve will, of course, require a profound shift in human consciousness which would devalue the role of economics in organizing lives and landscapes.

Such a shift in consciousness will not come easily in a world run by aggressive capitalists whose best interest is to prevent people from reaching the upper levels of Maslow's hierarchy of needs, where individuals would likely abandon crass consumerism for the benefits of self-actualization and aesthetic-cognitive values. Nor will the much more modest goal of environmental education be readily implemented in countries controlled by arrogant political and bureaucratic simpletons who appear to think that it is still in our best interest to live in 'God's own junkyard' while spending trillions of taxpayers' dollars on space programmes, supercolliders, and yet more weapons systems.

As has been recognized by generations of thinkers from Gautama Buddha to modern humanistic psychologists (Csikszentmihalyi 1993), the world will not be changed for the better in any fundamental way by coercion, legislation, or even top-down education. Rather, the texture of the future depends first on the myriad small, positive life-decisions made by millions of human individuals, and second, on the spiritual enterprise of small bands of visionaries who demonstrate alternative pathways by example.

At bottom, like much else in life, it comes down to a personal existential choice. The easy route is to succumb to conformist individualism, think of yourself as a consumer, shop till you drop, and degenerate into a minimal self (Lasch 1984), one of T.S. Eliot's 'stuffed men', so stuffed that they are hollow. This is, of course, a consummation devoutly wished upon us by those who organize the world apparently for our, but mostly for their, benefit (Don't worry; be happy). Alternatively, you can try to think and sense for yourself, even intuit and feel for yourself, embrace wild culture, and become a person, following your bliss. It's a tougher route.

In *Landscapes of the Mind* (1990) I echoed Jesus in calling for a more childlike (not childish) appreciation of earth and world. The reader of this book, I hope, will not hesitate to close it right now and go out to encounter a moral landscape on aesthetico-ethical terms, suffused perhaps with a sense of spirituality and feelings of attachment. Thus landscapes may not only delight us but also become our fields of care (Evernden 1985).

If the book is still open, you might end by, first, pondering the words of a young Alaskan Indian hunter (Nelson 1993: 225) to the effect that Western civilization 'is smart but not wise', and then, having absorbed that, return to Wordsworth ('The Tables Turned', vv. 7 and 8):

> Sweet is the lore which Nature brings;
> Our meddling intellect
> Mis-shapes the beauteous forms of things:
> We murder to dissect.

> Enough of Science and of Art:
> Close up those barren leaves;
> Come forth, and bring with you a heart
> That watches and receives.

BIBLIOGRAPHY

Abbey, E. (1977) *The Journey Home*. New York; Dutton.
—— (1979) *Abbey's Road*. New York; Dutton.
—— (1985) *The Monkey Wrench Gang*. Salt Lake City; Dream Garden Press.
Ackerman, D. (1990) *A Natural History of the Senses*. New York; Random House.
Acking, C.A. and R. Kuller (1973) Presentation and judgement of planned environment and the hypothesis of arousal; pp. 72–84 in W.F.E. Preiser (ed) *Environmental Design Research*. Stroudsburg, PA; Dowden, Hutchinson and Ross.
Acking, C.A. and G.J. Sorte (1973) How do we verbalise what we see? *Landscape Architecture* 64; 470–5.
Aiken, S.R. (1976) Toward landscape sensibility. *Landscape* 20; 20–8.
Aldous. T. (1972) *Battle for the Environment*. London; Fontana.
Aldridge, D. (1975) *Guide to Countryside Interpretation 1*. Edinburgh; HMSO.
Alexander, C., S. Ishikawa and M. Silverstein (1977) *A Pattern Language*. New York; Oxford University Press.
Allen, H. (1989) Landscape into art. *Manchester Guardian Weekly* 1 January.
Alsop, S. (1962) America the ugly. *Saturday Evening Post* 23 June; pp. 8–10.
Altman, I. and E.H. Zube (eds) (1990) *Public Places and Spaces*. New York; Plenum.
Amedeo, D. and R.A. York (1990) Indication of environmental schemata from thoughts about environments. *Journal of Environmental Psychology* 10; 219–53.
Anderson, L.M. and H.W. Schroeder (1983) Application of wildland scenic assessment methods to the urban landscape. *Landscape Planning* 10; 219–37.
Andresen, J.W. (1978) The greening of urban America. *American Forests* 84; 10–12, 56–61.
Appleton, J. (1975a) *The Experience of Landscape*. London; Wiley.
—— (1975b) Landscape evaluation: the theoretical vacuum. *Transactions of the Institute of British Geographers* 66; 120–3.
—— (ed.) (1976) *The Aesthetics of Landscape*. Didcot, UK; Rural Planning Services Ltd. Publication 7.
—— (1980) *David Linton's Contribution to Landscape Evaluation: A Critical Appraisal*. Birmingham; University of Birmingham, Department of Geography, Occasional Publication 13.
—— (1984) Prospects and refuges revisited. *Landscape Journal* 8; 91–103.
—— (1990) *The Symbolism of Habitat*. Seattle; University of Washington Press.
Appleyard, D. (1969) Why buildings are known. *Environment and Behavior* 1; 131–56.
Appleyard, D. and L. Fishman (1977) High-rise buildings versus San Francisco; pp. 81–100 in D.J. Conway (ed) *The Human Response to Tall Buildings*. Stroudsburg, PA; Dowden, Hutchinson and Ross.
Appleyard, D., K. Lynch and J.R. Meyer (1964) *The View from the Road*. Cambridge, MA; MIT Press.
Arnstein, S. (1969) A ladder of citizen participation. *Journal of the American Institute of Planners* 35; 216–24.
Arthur L.M. (1977) Predicting scenic beauty of forest environments: some empirical tests. *Forestry Science* 2; 151–60.
Arthur, L.M. and R.S. Boster (1976) *Measuring Scenic Beauty: A Selected Annotated Bibliography*. Fort Collins, Colorado; USA Forest Service General Technical Report RM-25.
Arthur, L.M., T.C. Daniel and R.S. Boster (1977) Scenic assessment: an overview. *Landscape Planning* 4; 109–29.

Artibise, A. and G.A. Stelter (eds) (1979) *The Usable Urban Past*. Toronto; Macmillan.

Astles, A. (1972) *The Evolution and Role of Historic and Architectural Preservation within the North American City*. Unpublished PhD thesis, Simon Fraser University, Vancouver.

Attoe, W. (1981) *Skylines: Understanding and Molding Urban Silhouettes*. New York; Wiley.

Avocat, C. (1982) Approche du paysage. *Revue de Géographie de Lyon* 57; 333–42.

Bachelard, G. (1964) *The Poetics of Space*. Boston; Beacon Press.

Back, K.W. and L.B. Bourque (1970) Can feelings be enumerated? *Behavioral Science* 15; 487–96.

Bacon, E. (1976) *Design of Cities*. New York; Penguin.

Bagley, M.D. (1975) Aesthetic assessment methodology; pp. 95–105 in L.E. Coate and P.A. Bonner (eds) *Regional Environmental Management*. Toronto; Wiley.

Bailly, A. (1977) *La perception de l'espace urbain*. Paris; Centre de recherche d' urbanisme.

Bailly, A., C. Raffestan and H. Raymond (1980) Les concepts du paysage: problématique et représentations. *L'Espace Géographique* 9; 277–86.

Balling, J.D. and J.H. Falk (1982) Development of visual preference for natural environments. *Environment and Behavior* 14; 5–28.

Bannerman, S. and P. Dearden (1985) Visual resource analysis in British Columbia; pp. 59–76 in M. Edgell (ed) *Geographical Research in the 1980s*. Victoria, BC; BC Geographical Series.

Barbour, I. (ed) (1972) *Earth Might be Fair: Reflections on Ethics, Religion and Ecology*. Englewood Cliffs, NJ; Prentice Hall.

———(1990) *Religion in an Age of Science*. San Francisco; Harper and Row.

Barrell, J. (1972) *The Idea of Landscape and the Sense of Place 1730–1840: An Approach to the Poetry of John Clare*. Cambridge; Cambridge University Press.

———(1980) *The Dark Side of the Landscape*. Cambridge; Cambridge University Press.

Baumgartner, R. (1984) *Die Visuelle Landschaft: Kartierung der Ressource Landschaft in den Colorado Rocky Mountains, USA*. Bern, Switzerland; Geographica Bernensia 22.

Beardsley, M.C. (1982) *The Aesthetic Point of View*. Ithaca, NY; Cornell University Press.

Bechtel, R.B. (1967) Human movement and architecture. *Trans-Action* 4; 53–6.

Beckett, P.H.T. (1974) The interaction between knowledge and aesthetic appreciation. *Landscape Research News* 1; 5–7.

Beer, A.R. (1990) *Environmental Planning for Site Development*. London; Spon, Chapman and Hall.

——— (1991) Urban design: The growing influence of environmental psychology. *Journal of Environmental Psychology* 11; 359–71.

Bell, C. and R. Bell (1969) *City Fathers*. London; Barrie and Rockliff.

Bennett, T. (1988) Museums and 'the people'; pp. 37–43 in R. Lumley (ed) *The Museum Time-Machine*. London; Murray.

Berezovoy, A. (1983) Salyut seedlings comfort spacemen. *The Guardian* 7 January.

Berger, J. (1972) *Ways of Seeing*. Harmondsworth, UK; Penguin.

——— (1980) *About Looking*. New York; Pantheon.

Berleant, A. (1992) *The Aesthetics of Environment*. Philadelphia; Temple University Press.

Berlyne, D.E. (1960) *Conflict Arousal and Curiosity*. New York; McGraw-Hill.

——— (1971) *Aesthetics and Psychology*. New York; Appleton-Century-Crofts.

——— (ed) (1974) *Studies in the New Experimental Aesthetics*. New York; Wiley.

Berman, M. (1982) *All That is Solid Melts into Air*. New York; Simon and Schuster.

Betjeman, J. (1980) *Collected Poems*. London; Murray.

Birkenhead, Lord. (1980) Introduction to J. Betjeman, *Collected Poems*. London; Murray.

Blake, P. (1964) *God's Own Junkyard*. New York; Holt, Rinehart and Winston.

Bloom, A. (1987) *The Closing of the American Mind*. New York; Simon and Schuster.

Bosselman, P., J. Flores and T. O'Hare (1983) *Sun and Light For Downtown San Francisco*. Berkeley, Calif.; University of California Environmental Simulation Laboratory.

Bouman, J. (1979) Perception of environmental quality: some basic sensory conditions. *Urban Ecology* 4; 97–101.

Bourassa, S.C. (1991) *The Aesthetics of Landscape*. London; Belhaven.

Brace P. (1980) Urban aesthetics and the courts. *The Urban Lawyer* 12; 151–6.

Brancher, D.M. (1969) Critique of K.D. Fines' landscape evaluation. *Regional Studies* 3; 91–2.

Brandon, P.F. and R.N. Millman (eds) (1978) *Historic Landscapes: Identification, Recording, Management*. London; Polytechnic of North London Occasional Paper.

Brebner, J. (1982) *Environmental Psychology in Building Design*. Barking, UK; Applied Science Publishers.

Broughton, R. (1972) Aesthetics and environmental law: decisions and values. *Land and Water Law Review* 7; 451–500.

Buckley, W.F. (1966) The politics of beauty. *Esquire* July, 50–6.

Buhyoff, G.J. and J.D. Wellman (1979) Seasonality bias in landscape preference research. *Leisure Sciences* 2; 181–90.

Buhyoff, G.J., J.D. Wellman, H. Harvey and R.A. Fraser (1978) Landscape architects' interpretations of people's landscape preferences. *Journal of Environmental Management* 6; 255–62.

Bunkse, E.V. (1978) Commoner attitudes toward landscape and nature. *Annals of the Association of American Geographers* 68; 551–66.

Burchard, J.E. (1957) The urban aesthetic. *Annals of the Academy of Political and Social Science* 314; 112–22.

Burke, E. (1909) *On Taste: The Sublime and Beautiful*. New York; Collier.

Burke, G. (1976) *Townscapes*. Harmondsworth; Penguin.

Burton, L.M. (1981) *A Critical Analysis and Review of the Research on Outward Bound and Related Programs*. Unpublished PhD thesis, Rutgers University, New Brunswick NJ.

Buttimer, A. (1993) *Geography and the Human Spirit*. Baltimore; Johns Hopkins University Press.

Byrom, C. (1974) Perceptions of environmental quality on housing estates; pp. 165–77 in J.T. Coppock and C.B. Wilson (eds) *Environmental Quality*. Edinburgh; Scottish Academic Press.

Canter, D. (1977) *The Psychology of Place*. London; Architectural Press.

Canter, D. and R. Thorne (1972) Attitudes to housing: a cross-cultural comparison. *Environment and Behavior* 4; 3–32.

Carls, E.G. (1979) Coastal recreation: esthetics and ethics. *Coastal Zone Management Journal* 5; 119–30.

Carlson, A. (1976) Environmental aesthetics and the dilemma of aesthetic education. *Journal of Aesthetic Education* 10; 69–82.

———— (1977) On the possibility of quantifying scenic beauty. *Landscape Planning* 4; 131–72.

———— (1979) Appreciation and the natural environment. *Journal of Aesthetics and Art Criticism* 7; 267–75.

———— (1990) Whose vision? whose meaning? whose values? in P. Groth (ed) *Vision, Culture and Landscape: Working Paper from the Berkeley Symposium on Cultural Landscape Interpretation*. Berkeley, Calif.; University of California Department of Landscape Architecture.

Carpenter, E. (1973) *Eskimo Realities*. New York; Holt, Rinehart and Winston.

Carr, E. (1966) *The Book of Small*. Toronto; Clarke, Irwin.

———— (1978) *Hundreds and Thousands*. Toronto; Clarke, Irwin.

Carruth, D.B. (1977) Assessing scenic quality. *Landscape* 22; 31–4.

Chalmers, D. (1978) Environmental aesthetics: concepts and methods; pp. 23–48 in *Proceedings of a Workshop on Environmental Perception*. University of Otago, New Zealand.

Chandler, H.P. (1922) The attitude of the law toward beauty. *American Bar Association Journal* 8; 470–4.

Charles, Prince of Wales (1989) *A Vision of Britain*. London; Doubleday.

Cherulnik, P.D. (1993) *Applications of Environment–Behavior Research: Case Studies and Analysis*. Cambridge; Cambridge University Press.

Chevalier, L. (1993) *The Assassination of Paris*. Chicago; University of Chicago Press.

Child, I.L. (1962) Personal preferences as an expression of aesthetic sensitivity. *Journal of Personality* 30; 496–512.

Child, I.L. and S. Iwao (1968) Personality and aesthetic sensitivity. *Journal of Personality and Social Psychology* 8; 308–12.

Clamp, P. (1981) The landscape evaluation controversy. *Landscape Research* 6; 13–15.

Clamp. P. and M. Powell (1982) Prospect–refuge theory under test. *Landscape Research* 7; 7–8.

Clark, K. (1956) *Landscape into Art*. Harmondsworth, UK; Penguin.

Classen, C. (1993) *World of Sense: Exploring the Senses in History and across Cultures*. London; Routledge.

Clout, H. (1972) *Rural Geography*. Oxford; Pergamon.

Cobb, R. (1975) *A Sense of Place*. London; Duckworth.

Cobham, R. (1984) Blenheim: The art and management of landscape restoration. *Landscape Research* 9; 4–14.

Coe, R. (1984) *When the Grass was Taller: Autobiography and the Experience of Childhood*. New Haven, Conn.; Yale University Press.

Coeterier, J.F. (1983) A photo validity test. *Journal of Environmental Psychology* 3; 315–23.

Cole, B. (1980) *Sienese Painting from its Origins to the Fifteenth Century*. New York; Harper.

Cole, M. and S. Scribner (1974) *Culture: A Psychological Introduction*. New York; Wiley.

Collot, M. (1986) 'Points de vue sur la perception des paysages, *L'Espace Géographique* 3; 211–7.

Cooper Marcus, C. (1978) Remembrance of landscapes past. *Landscape* 22; 34–43.

——— (1979) *Environmental Autobiography*. Berkeley, Calif.; University of California Institute of Urban and Regional Development Working Paper 301.

Cooper Marcus, C. and W. Sarkissian (1986) *Housing as if People Mattered*. Berkeley, Calif.; University of California Press.

Corbin, A. (1986) *Le Miasme et la Jonquille: L'Odorat et L'Imaginaire Social XVIIIe–XIXe Siècles*. Paris; Flammarion.

——— (1994) *Les Cloches de la Terre*. Paris; Albin Michel.

Cornish, V. (1931) *The Poetic Impression of Natural Scenery*. London; Sifton Praed.

——— (1935) *Scenery and the Sense of Sight*. Cambridge; Cambridge University Press.

Cornwell, R. (1989) MOMA's ego builder. *New Art Examiner* 16; 36–41.

Correy, A. (1977) Land values: changing attitudes towards the Australian environment. *Planning Outlook* 30; 14–27.

Cosgrove, D. (1994) Contested global visions: one-world, whole-earth and the Apollo space photographs. *Annals of the Association of American Geographers* 84; 270–94.

Coventry–Solihull–Warwickshire Sub-Regional Planning Group (1971) *A Strategy for the Sub-Region*. Coventry; C–S–W SRP Supplementary Report (Countryside) 5.

Craik, K.H. (1968) The comprehension of the everyday physical environment. *Journal of the American Institute of Planners* 34; 29–37.

——— (1970) *A System of Landscape Dimensions*. Berkeley, Calif.; Institute for Personality Assessment and Research, Report to Resources for the Future Inc.

——— (1972a) Appraising the objectivity of landscape dimensions; pp. 292–346 in J.V. Krutilla (ed) *Natural Environments: Studies in Theoretical and Applied Analysis*. Baltimore; Johns Hopkins University Press.

——— (1972b) Psychological factors in landscape appraisal. *Environment and Behavior* 4; 259–65.

——— (1976) The personality research paradigm in environmental psychology; pp. 55–79 in S. Wapner, S.B. Cohen and B. Kaplan (eds) *Experiencing the Environment*. New York; Plenum.

Crofts, R.S. (1975) The landscape component approach to landscape evaluation. *Transactions, Institute of British Geographers* 66; 124–9.

Crofts, R.S. and R.V. Cooke (1974) *Landscape Evaluation: A Comparison of Techniques*. London; University College London Department of Geography Occasional Paper 25.

Crosby, T. (1967) *Architecture: City Sense*. London; Studio Vista.

——— (1973) *How to Play the Environment Game*. London; Penguin.

Crowe, S. (1971) *Tomorrow's Landscape*. London; Architectural Press.

Csikszentmihalyi, M. (1993) *The Evolving Self*. New York; Harper Collins.

Cullen, G. (1961) *The Concise Townscape*. London; Architectural Press.

Cunliffe, M. (1969) The eye and the mind's heart: The aesthetics of townscape. *Journal of the Town Planning Institute* 55; 162–7.

Cunningham, F.F. (1975) The human eye and the landscape. *Landscape* 20; 14–19.

Cupitt, D. (1984) *The Sea of Faith*. London; British Broadcasting Corporation.

Danford, S. and E.P. Willems (1975) Subjective responses to architectural displays: a question of validity. *Environment and Behavior* 7; 486–516.

Daniel, T.C. and R.S. Boster (1976) *Measuring Landscape Aesthetics: The Scenic Beauty Estimation Method*. Fort Collins, Colorado; USDA Forest Service Research Paper RM–167.

Daniel, T.C. and J. Vining (1983) Methodological issues in the assessment of landscape quality; pp. 39–84 in I. Altman and J. Wohlwill (eds) *Behavior and the Natural Environment*. New York; Plenum.

Daniel, T.C., R.S. Boster and L. Wheeler (1977) Mapping scenic beauty. *Leisure Sciences* 1; 31–52.

Daniel, T.C., L. Wheeler, R.S. Boster and P. Best (1973) Quantitative evaluation of landscapes: an application of signal detection analysis to forest management alternatives. *Man–Environment Systems* 3; 330–44.

Davidson, J. and C. Tayler (1970) Herefordshire Countryside Plan 1970. *Recreation News Supplement* 2; 43–4.

Davies, R. (1980) *Tempest-tost*. Toronto; Penguin.

Dearden, P. (1980a) Landscape assessment: The last decade. *Canadian Geographer* 24; 316–25.

—— (1980b) A statistical method for the assessment of visual landscape quality for land use planning purposes. *Journal of Environmental Management* 9; 51–68.

—— (1981) Public participation and scenic quality analysis. *Landscape Planning* 8; 3–19.

—— (1982) Some comments in reply to D.L. Jacques. *Journal of Environmental Management* 14; 237–8.

—— (1983) Forest harvesting and landscape assessment techniques in British Columbia, Canada. *Landscape Planning* 10; 239–53.

—— (1984) Factors influencing landscape preferences: an empirical investigation. *Landscape Planning* 11; 293–306.

—— (1987) Consensus and a theoretical framework for landscape evaluation. *Journal of Environmental Management* 34; 267–78.

—— (1989) Societal landscape preferences: a pyramid of influences; pp. 41–63 in P. Dearden and B. Sadler (eds) *Landscape Evaluation*. Victoria, BC; Western Geographical Series.

Dearden, P. and L. Rosenblood (1980) Some observations on multivariate techniques in landscape evaluation. *Regional Studies* 14; 99–110.

Dearden P. and B. Sadler (eds) (1989) *Landscape Evaluation: Approaches and Applications*. Victoria, BC; Western Geographical Series.

Debord, G. (1967) *The Society of the Spectacle*. Detroit; Red and Black.

De Quincey, T. (1839) *Murder Considered as one of the Fine Arts*. Edinburgh; Blackwood.

Dickie, G. (1962) Is psychology relevant to aesthetics? *Philosophical Review* 71; 285–303.

—— (1971) *Aesthetics*. Indianapolis; Pegasus.

Dixon, J.M. (1988) Progressive Architecture reader poll design preferences. *Progressive Architecture* 69; 15–17.

Dobby, A. (1978) *Conservation and Planning*. London; Hutchinson.

Dornbusch, D.M. and P.M. Gelb (1977a) High-rise visual impact; pp. 101–11 in D.J. Conway (ed) *Human Response to Tall Buildings*. Stroudsburg, PA; Dowden, Hutchinson and Ross.

Dornbusch, D.M. and P.M. Gelb (1977b) High-rise impacts on the use of parks and plazas; pp. 112–30 in D.J. Conway (ed) *Human Response to Tall Buildings*. Stroudsburg, PA; Dowden, Hutchinson and Ross.

Dunham, K. (1994) *Beautiful British Columbia*? Unpublished MA thesis, University of Victoria, BC, Department of Geography.

Dunn, M.C. (1974a) *Landscape Evaluation Techniques: A Review and Appraisal of the Literature*. Birmingham; Centre for Urban and Regional Studies Working Paper 4.

—— (1974b) Landscape evaluation: a further perspective. *The Planner: Journal of the Royal Town Planning Institute* 60; 935–6.

—— (1976) Landscape with photographs: testing the preference approach to landscape evaluation. *Journal of Environmental Management* 4; 15–26.

Eichler, E. and M. Kaplan (1967) *The Community Builders*. Berkeley, Calif.; University of California Press.

Eliot, T.S. (1948) *Notes towards the Definition of Culture*. London; Faber.

—— (1974) *Collected Poems 1909–62*. London; Faber.

Ellul, J. (1981) Socialism yes . . . an interview. *Manchester Guardian Weekly* 15 November.

Erickson, K.A. (1977) Ceremonial landscapes of the American West. *Landscape* 22; 39–47.

European Community (1990) *The Green Paper on the Urban Environment*. Brussels; Commission of the European Community.

Evans, G.W. and K.W. Wood (1980) Assessment of environmental aesthetics in scenic corridors. *Environment and Behavior* 12; 255–73.

Evans, G.W. and E.H. Zube (1975) Information processing. *Man–Environment Systems* 5; 61–2.

Evans, P. (1995) Wild at heart. *The Geographical Magazine* 67; 27–9.

Evernden, N. (1985) *The Natural Alien*. Toronto; University of Toronto Press.

Fabos, J.G. (1974) Putting numbers on qualities: The rising landscape assessors. *Landscape Architecture* 64; 164–5.

Fabos, J.G., W.G. Hendrix and C.M. Greene (1975) Visual and cultural components of the landscape resource assessment model of the METLAND study; pp 319–43 in E.H. Zube, R.O. Brush and J.G. Fabos (eds) *Landscape Assessment*. Stroudsburg, PA; Dowden, Hutchinson and Ross.

Faludy, G. (1988) *Notes from the Rainforest*. Willowdale, Ontario; Hounslow Press.

Fenton, D.M. and J.P. Reser (1988) The assessment of landscape quality: an integrative approach; pp. 108–19 in J.L. Nasar (ed) *Environmental Aesthetics*. Cambridge; Cambridge University Press.

Fines, K.D. (1968) Landscape evaluation: a research project in East Sussex. *Regional Studies* 2; 41–55.

Flaschbart, P.G. and G.L. Peterson (1973) Dynamics of preference for visual attributes of housing environments; pp. 98–106 in W.F.E. Preiser (ed) *Environmental Design Research IV*. Stroudsburg, PA; Dowden, Hutchinson and Ross.

Foote, K.E. (1983) *Colour in Public Spaces: Toward a Communication-based Theory of the Urban Built Environment*. Chicago; University of Chicago Department of Geography Research Paper 205.

Ford, L.R. (1974) Historic preservation and the sense of place. *Growth and Change* 5; 33–7.

⸺ (1979) Urban preservation and the geography of the city in the USA. *Progress in Human Geography* 3; 211–38.

⸺ (1994) *Cities and Buildings*. Baltimore; Johns Hopkins University Press.

Foreman, D. (ed) (1985) *Ecodefense: A Field Guide to Monkeywrenching*. No place; no publisher.

Forster, E.M. (1910) *Howard's End*. New York; Knopf.

Foster, H. (1985) *Recoding: Art, Spectacle, Cultural Politics*. Port Townsend, WA; Bay Press.

Francis, C. and C. Cooper Marcus (1990) *People Places: Design Guidelines for Urban Open Space*. New York; Van Nostrand Reinhold.

Francis, M. (1987) Urban open spaces; pp. 71–106 in E.H. Zube and G.T. Moore (eds) *Advances in Environment, Behavior, and Design 1*. New York; Plenum.

Francis, M. and R.T. Hester (eds) (1988) *The Meaning of Gardens*. Cambridge, MA; MIT Press.

Frye, N. (1983) *The Great Code*. Toronto; University of Toronto Press.

⸺ (1991) *The Double Vision: Language and Meaning in Religion*. Toronto; University of Toronto Press.

Fuller, P. (1988) The geography of mother nature; pp. 11–31 in D. Cosgrove and S. Daniels (eds) *The Iconography of Landscape*. Cambridge; Cambridge University Press.

Galbraith, J.K. (1966) Economics and environment. *American Institute of Architecture Journal* 46; 55–8.

Gardiner, J. (1971) Cruel aesthetics. *The Architect* 1; 35–9.

Garling, T. (1976) The structural analysis of environmental perception and cognition: a multidimensional scaling approach. *Environment and Behavior* 8; 385–415.

Gaunt, W. (1988) *The Aesthetic Adventure*. London; Cardinal.

Gelb, P.M. (1977) High-rise impact on city and neighborhood livability; pp. 131–47 in D.J. Conway (ed) *The Human Response to Tall Buildings*. Stroudsburg, PA; Dowden, Hutchinson and Ross.

Gibson, E. (1989) Traditions of landscape aesthetics; pp. 23–38 in P. Dearden and B. Sadler (eds) *Landscape Evaluation*. Victoria, BC; Western Geographical Series.

Gibson, J.J. (1966) *The Senses Considered as a Perceptual System*. Boston; Houghton Mifflin.

Gifford, R. (1987) *Environmental Psychology*. Boston; Allyn and Bacon.

Gilpin, W. (1786) *Observations Relative Chiefly to Picturesque Beauty, Made in the Year 1772 on Several Parts of England*. London; Blamire.

Gilg, A.W. (1974a) The objectivity of Linton type methods of assessing scenery as a natural resource. *Regional Studies* 9; 181–91.

⸺ (1974b) A critique of Linton's method of assessing scenery as a natural resource. *Scottish Geographical Magazine* 90; 125–9.

⸺ (1976) Assessing scenery as a natural resource: causes of variation in Linton's method. *Scottish Geographical Magazine* 92; 41–9.

Girouard, M. (1985) *Cities and People*. New Haven, Conn.; Yale University Press.

Gloor, P. (1978) Inputs and outputs of the amygdala; pp. 36–41 in K.E. Livingston and O. Hornykiewica (eds) *Limbic Mechanisms*. New York; Plenum.

Gold, J. (1980) *An Introduction to Behavioural Geography*. Oxford; Oxford University Press.

Gold, J. and J. Burgess (eds) (1982) *Valued Environments*. London; Allen and Unwin.

Gold, M. (1984) A history of nature; pp. 12–33 in D. Massey and J. Allen (eds) *Geography Matters!* Cambridge; Cambridge University Press.

Goodey, B. (1977) *Interpreting the Conserved Environment*. Oxford; Oxford Polytechnic, Department of Town Planning, Working Paper 29.

⸺ (1978a) *Sensing the Environment*. London; Town and Country Planning Association.

⸺ (1978b) Where we're at: interpreting the urban environment. *Urban Design Forum* 1; 28–34.

⸺ (1982) Values in place: interpretations and implications from Bedford; pp. 10–34 in J.R. Gold and J. Burgess (eds) *Valued Environments*. London; Allen and Unwin.

Graber, L. (1976) *Wilderness as Sacred Space*. Washington, DC; Association of American Geographers Monograph 8.

Greenbie, B. (1982) The landscape of social symbols. *Landscape Research* 7; 2–6.

Green, G.W. (ed) (1856) *The Works of James Addison*. New York; Scribner's.

Groat, L. (1982) Meaning in Post-Modern architecture: an examination using the multiple sorting task. *Journal of Environmental Psychology* 2; 3–22.

Groat, L. and C. Després (1991) The significance of architectural theory for environmental design research; pp. 3–51 in E. Zube and G.T. Moore (eds) *Advances in Environment, Behavior, and Design 3*. New York; Plenum.

Grosjean, G. (1986) *Asthetische Bewertung ländlicher Räume am Beispell von Grindelwald*. Bern, Switzerland; Schlussbericht zum Schweizerischen MAB–Programm 20.

Gussow, A. (1977) In the matter of scenic beauty. *Landscape* 21; 26–35.

Hackney, S. (1995) Lasting values in a disposable world. *Association of American Geographers Newsletter* 30; 21–31.

Hall, D.R. (1982) Valued environments and the planning process; pp. 172–88 in J.R. Gold and J. Burgess (eds) *Valued Environments*. London; Allen and Unwin.

Halprin, L. (1969) *The RSVP Cycles: Creative Processes in the Urban Environment*. New York: Braziller.

——— (1970) *Everett*. Report by L. Halprin Associates to the City of Everett, Washington State.

Hamill, L. (1989) On the persistence of error in landscape aesthetics; pp. 197–206 in P. Dearden and B. Sadler (eds) *Landscape Evaluation*. Victoria, BC; Western Geographical Series.

Harrington, D.L. and W.F.E. Prieser (1980) Layperson assessments of visual and aesthetic quality; pp. 219–31 in R.Thorne and D. Arden (eds) *People and the Man-made Environment*. Sydney; University of Sydney Press.

Harry, J., R. Gale and J. Hendee (1969) Conservation: an upper-middle class social movement. *Journal of Leisure Research* 1; 219–37.

Harvey, D. (1985) *The Urbanization of Capital*. Baltimore; Johns Hopkins University Press.

Haskell, D. (1965) The drive for beauty and the popular taste. *Landscape* 15; 2–5.

Healey, M. (1989) The right tone, the right time. Letter to the editor, *Manchester Guardian Weekly* 29 January.

Heath, T.F. (1974) Should we tell the children about aesthetics, or should we let them find out in the street? pp. 179–83 in D. Canter and T. Lee (eds) *Psychology and the Built Environment*. London; Architectural Press.

Heerwagen, J.H. and G.H. Orians (1993) Humans, habitats, and aesthetics; pp. 138–72 in S.R. Kellert and E.O. Wilson (eds) *The Biophilia Hypothesis*. Washington DC; Island Press.

Helphand, K.I. (1979) Environmental autobiography. Paper given at The International Conference on Environmental Psychology, University of Surrey, Guildford, UK.

Hendrix, W.G. and J.G. Fabos (1975) Visual land use compatibility as a significant contributor to visual resource quality. *International Journal of Environmental Studies* 8; 21–8.

Hepburn, R.W. (1963) Aesthetic appreciation of nature. *British Journal of Aesthetics* 3; 195–209.

——— (1968) Aesthetic appreciation in nature; pp. 49–66 in H. Osborn (ed) *Aesthetics in the Modern World*. London; Thames and Hudson.

Hepworth J. and G. McNamee (1985) *Resist Much, Obey Little: Some Notes on Edward Abbey*. Salt Lake City; Dream Garden Press.

Herbert, D.T. (1989) Does interpretation help? pp. 191–230 in D.T. Herbert, R.C. Price and C.J. Thomas (eds) *Heritage Sites: Strategies for Marketing and Development*. Aldershot, UK; Avebury.

Herzog, T.R. and P. Bosley (1992) Tranquillity and preference as affective qualities of natural environments. *Journal of Environmental Psychology* 12; 115–29.

Herzog, T.R., S. Kaplan and R. Kaplan (1976) The prediction of preference for familiar urban places. *Environment and Behavior* 8; 627–45.

Hester, R. (1979) A womb with a view: how spatial nostalgia affects the designer. *Landscape Architecture* September, 475–82.

——— (1993) Sacred structures and everyday life; pp. 271–97 in D. Seamon (ed) *Dwelling, Seeing, and Designing*. Albany, NY; SUNY Press.

Hewison, R. (1987) *The Heritage Industry: Britain in a Climate of Decline*. London; Methuen.

Heyligers, P.C. (1981) Prospect–refuge symbolism in dune landscapes. *Landscape Research* 6; 7–11.

Hibbitts, B.J. (1992) Coming to our senses: communication and legal expression in performance cultures. *Emory Law Journal* 41; 873–960.

Hildebrand, G. (1991) *The Wright Space*. Seattle; University of Washington Press.
Hillier, B., J. Musgrave and P. O'Sullivan (1972) Knowledge and design; pp. 29.3.1–29.3.9 in W.J. Mitchell (ed) *Environmental Design: Research and Practice*. Los Angeles; University of California Press.
Hirsch, E.D. (1987) *Cultural Literacy*. Boston; Houghton Mifflin.
Hospers, J. (1946) *Meaning and Truth in the Arts*. Chapel Hill, NC; University of North Carolina Press.
Hough, M. (1990) *Out of Place: Restoring Identity to the Regional Landscape*. New Haven; Yale University Press.
Howard, E. (1898) *Garden Cities of Tomorrow*. London; Sonnenschein.
Howes, D. (ed) (1991) *The Varieties of Sensory Experience*. Toronto; University of Toronto Press.
Hudson, L. (1966) *Contrary Imaginations*. New York; Schocken Books.
Hull, R.B. and G.R.B. Revell (1989) Cross-cultural comparison of landscape scenic beauty evaluations: a case-study in Bali. *Journal of Environmental Psychology* 9; 172–92.
Hummel, M. (ed) (1989) *Endangered Spaces: The Future for Canada's Wilderness*. Toronto; Key Porter.
Hummon, D.M. (1990) *Common Places*. New York; State University of New York Press.
Hussey, C. (1929) *The Picturesque: Studies in a Point of View*. New York; Putnam.
Huxtable, A.L. (1970) *Will They Ever Finish Bruckner Boulevard?* New York; Macmillan.
—— (1986) *Goodbye History: Hello Hamburger*. Washington DC; Preservation Press.
Im, S-B. (1984) Visual preferences in enclosed urban spaces. *Environment and Behavior* 16; 235–62.
Itami, R.M. (1989) Scenic perception: research and application in US visual management systems; pp. 211–41 in P. Dearden and B. Sadler (eds) *Landscape Evaluation*. Victoria, BC; Western Geographical Series.
Jackson, J.B. (1959) The imitation of nature. *Landscape* 9; 9–12.
—— (1964) Limited access. *Landscape* 14; 18–23.
Jackson, R.H., L.E. Hudman and J.L. England (1978) Assessment of the environmental impact of high voltage power transmission lines. *Journal of Environmental Management* 6; 153–70.
Jacobs, J. (1961) *The Death and Life of Great American Cities*. New York; Random House.
Jacobs, P. (1975) The landscape image: current approaches to the visual analysis of the landscape. *Town Planning Review* 46; 127–50.
Jakle, J.A. (1987) *The Visual Elements of Landscape*. Amherst, MA; University of Massachusetts Press.
Jakle, J.A. and R.W. Travis (1980) *Preserving Past Landscapes: Historic Preservation as Environmental Management*. Stroudsburg, PA; Dowden, Hutchinson and Ross.
Janssens, J. (1984) *Looking at Buildings*. Lund, Sweden; Lund Institute of Technology, Department of Theoretical and Applied Aesthetics.
Jencks, C. (1978) *The Language of Post-Modern Architecture*. New York; Rizzoli.
—— (1988) Deconstruction: The pleasure of abuse. *Architectural Design* 58; 3–4.
Jones, J.C. and D.G. Thornley (eds) (1963) *Conference on Design Methods*. New York; Macmillan.
Jones, M. (1991) The elusive reality of landscape: concepts and approaches in landscape research. *Norsk Geografisk Tidsskrift* 45; 229–44.
Jungst, P. and O. Meder (1986) *Towards a Grammar of Landscape: The Relationship between Scene and Space*. Kassel, Germany; Kasseler Schriften zur Geographie und Planung.
Kaplan, R. (1973a) Some psychological benefits of gardening. *Environment and Behavior* 5; 145–52.
—— (1973b) Predictors of environmental preference: designers and clients; pp. 265–74 in W.F.E. Preiser (ed) *Environmental Design Research*. Stroudsburg, PA; Dowden, Hutchinson and Ross.
—— (1975) Some methods and strategies in the prediction of preference; pp. 118–29 in E.H. Zube, R.O Brush and J.G. Fabos (eds) *Landscape Assessment*. Stroudsburg, PA; Dowden, Hutchinson and Ross.
—— (1977a) Preference and everyday nature: method and application; pp. 235–50 in D. Stokols (ed) *Perspectives on Environment and Behavior*. New York; Plenum.
—— (1977b) Patterns of environmental preference. *Environment and Behavior* 9; 195–216.
—— (1978a) Participation in environmental design: some considerations and a case-study; pp. 427–38 in S. Kaplan and R. Kaplan (eds) *Humanscape*. Belmont, Calif.; Duxbury.

Kaplan, R. (1978b) The green experience; pp. 186–93 in S. Kaplan and R. Kaplan (eds) *Humanscape*. Belmont, Calif.; Duxbury.

Kaplan, R. and E.J. Herbert (1988) Familiarity and preference: a cross-cultural analysis; pp. 379–89 in J.L. Nasar (ed) *Environmental Aesthetics*. Cambridge; Cambridge University Press.

Kaplan, R. and S. Kaplan (1989) *The Experience of Nature*. New York; Cambridge University Press.

Kaplan, S. (1979) Perception and landscape: conceptions and misconceptions; pp. 241–8 in G.H. Elsner and R.C. Smardon (eds) *Proceedings of 'Our National Landscape' Conference*. Berkeley, Calif.; USDA Forest Service General Technical Report PSW–35.

—— (1983) A model of person–environment compatibility. *Environment and Behavior* 15; 311–32.

—— (1987) Aesthetics, affect, and cognition: environmental preferences from an evolutionary perspective. *Environment and Behavior* 19; 3–32.

Kaplan, S. and R. Kaplan (1982) *Cognition and Environment*. New York; Praeger.

Kaplan, S. and J.F. Talbot (1983) Psychological benefits of a wilderness experience; pp. 73–86 in I. Altman and J. Wohlwill (eds) *Behavior in the Natural Environment*. New York; Plenum.

Kaplan, S., R. Kaplan and J.S. Wendt (1972) Rated preference and complexity for natural and urban visual material. *Perception and Psychophysics* 12; 354–6.

Katcher, A. and G. Wilkins (1993) Dialogue with animals: its nature and culture; pp. 173–200 in S.R. Kellert and E.O. Wilson (eds) *The Biophilia Hypothesis*. Washington DC; Island Press.

Kay, J. (1989) Human dominion over nature in the Hebrew Bible. *Annals of the Association of American Geographers* 79; 214–32.

Kellert, S.R. and E.O. Wilson (eds) (1993) *The Biophilia Hypothesis*. Washington DC; Island Press.

Kluckner, D. (1991) *Paving Paradise*. Vancouver, BC; Whitecap Books.

Koberg, D. and J. Bagnall (1974) *The Universal Traveller*. Los Altos, Calif.; Kaufmann.

Kuhn, F. (1968) Research in human space. *Ekistics* 25; 395–8.

Kunstler, J.H. (1993) *The Geography of Nowhere*. New York; Simon and Schuster.

Kurtz, J. (1971) And now a word from the users. *Design and Environment* 2; 41–8.

—— (1973) *Wasteland: Building the American Dream*. New York; Praeger.

Kutcher, A. (1972) *The New Jerusalem: Planning and Politics*. London; Thames and Hudson.

Ladd, F. (1977) Residential history. *Landscape* 21; 15–20.

Lang, J. (1988) Symbolic aesthetics in architecture: toward a research agenda; pp. 11–26 in J.L. Nasar (ed) *Environmental Aesthetics*. Cambridge; Cambridge University Press.

—— (1991) Design theory from an environment and behavior perspective; pp. 53–102 in E. Zube and G.T. Moore (eds) *Advances in Environment, Behavior, and Design 3*. New York; Plenum.

Larkin, P. (1988) *Collected Poems*. London; Faber.

Lasch, C. (1984) *The Minimal Self*. New York; Norton.

Laughlin, N.A. (1984) *Attitudes of Landscape Architects in the USDA Forest Service toward the Visual Management System*. Unpublished MLA thesis, School of Renewable Natural Resources, University of Arizona, Tucson.

Le Corbusier (1923) *Towards a New Architecture*. New York; Praeger.

Le Guérer, A. (1988) *Les Pouvoirs de l'Odeur*. Paris; Bourin.

Leopold, L.B. (1969a) *Quantitative Comparison of Some Aesthetic Factors among Rivers*. Washington DC; Geological Survey Circular 620.

—— (1969b) Landscape aesthetics. *Natural History* 78; 36–45.

Levy-Leboyer, C. (1977) *Étude Psychologique du Cadre de Vie*. Paris; Editions du CNRS.

Lewis, C.A. (1979) Healing in the urban environment: a person/plant viewpoint. *Journal of the American Institute of Planners* 45; 330–8.

Lewis, P.F. (1970) When cleanliness is not enough. *Public Management* 51; 8–11.

—— (1979) Axioms for reading the landscape; pp.11–32 in D.W. Meinig (ed) *The Interpretation of Ordinary Landscapes*. New York; Oxford University Press.

Lewis, P.F., D. Lowenthal and Y.-F. Tuan (1973) *Visual Blight in America*. Washington, DC; Association of American Geographers, College Geography Resource Paper 23.

Liddle, M.J. (1976) An approach to objective collection and analysis of data for comparison of landscape character. *Regional Studies* 10; 173–81.

Ligue Suisse Pour la Protection de la Nature (1979) *Inventaire des Paysages et des Sites Naturels d'Importance Nationale qui méritent Protection*. Basel; Schweizerischer Bund für Naturschutz.

Linton, D.L. (1968) The assessment of scenery as a natural resource. *Scottish Geographical Magazine* 84; 218–38.

Little, B.R. (1976) The personality research paradigm in environmental psychology; pp. 55–80 in S. Wapner, S.B. Cohen and B. Kaplan (eds) *Experiencing the Environment*. New York; Plenum.

Litton, R.B. (1974) Visual vulnerability of forest. *Journal of Forestry* 72; 394–7.

—— (1977) River landscape quality and its assessment; pp. 46–54 in USDA Forest Service (ed) *Proceedings of the River Recreation Management and Research Symposium*. Minneapolis; USDA Forestry Service. General Technical Report NC–28.

—— (1982) Visual assessment of natural landscapes; pp.97–115 in B. Sadler and A. Carlson (eds) *Environmental Aesthetics: Essays in Interpretation*. Victoria, BC; Western Geographical Series.

Locasso, R.M. (1988) The influence of a beautiful versus an ugly room on ratings of photographs of human faces: a replication of Maslow and Mintz; pp. 134–43 in J.L. Nasar (ed) *Environmental Aesthetics*. Cambridge; Cambridge University Press.

Lottman, H. (1976) *How Cities are Saved*. New York; Universe Books.

Lovelock, J.E. (1979) *Gaia: A New Look at Life on Earth*. Oxford; Oxford University Press.

Lowe, B. (1977) Environmental values. *Built Environment Quarterly* 3; 79–82.

Lowenthal, D. (ed) (1967) *Environmental Perception and Behaviour*. Chicago; Department of Geography, University of Chicago.

—— (1968) The American scene. *Geographical Review* 58; 61–8.

—— (1978a) *Finding Valued Landscapes*. Toronto; University of Toronto, Department of Geography Environmental Perception Working Paper 4.

—— (1978b) Finding Valued Landscapes. *Progress in Human Geography* 2; 373–418.

—— (1979) Environmental perception: preserving the past. *Progress in Human Geography* 3; 550–9.

—— (1980) *Boats Against the Current: A Study of How We Use the Past*. Cambridge; Cambridge University Press.

—— (1985) *The Past is a Foreign Country*. Cambridge; Cambridge University Press.

Lowenthal, D. and M. Binney (1981) *Our Past Before Us: Why Do We Save It?* London; Temple Smith.

Lowenthal, D. and H. Prince (1964) The English landscape. *Geographical Review* 54; 309–46.

—— (1965) English landscape tastes. *Geographical Review* 55; 186–222.

Lowry, M. (1970) *October Ferry to Gabriola*. New York; Plume.

Lozano, E. (1974) Visual needs in the environment. *Town Planning Review* 43; 351–74.

Lucas, B. (1993) Unlocking the landscape. *Landscape Design* 221; 11–17.

Lucas, R.C. (1964) Wilderness perception and use: the example of the Boundary Waters Canoe Area. *Natural Resources Journal* 3; 394–411.

—— (1970a) *User evaluation of campgrounds in two Michigan National Forests*. St. Paul, Minnesota; USDA Forest Service Research Paper NC–44.

—— (1970b) User concepts of wilderness and their implications for resource management; pp. 297–302 in H. Proshansky, W. Ittelson, and L. Rivlin (eds) *Environmental Psychology*. New York; Holt, Rinehart and Winston.

Lutz, C.A., and J.L. Collins (1993) *Reading National Geographic*. Chicago; University of Chicago Press.

Lynch, K. (1959) A walk around the block. *Landscape* 8; 24–34.

—— (1960) *The Image of the City*. Cambridge, MA; MIT Press.

—— (1972) *What Time is this Place?* Cambridge, MA; MIT Press.

—— (1976) *Managing the Sense of a Region*. Cambridge, MA; MIT Press.

—— (1981) *Good City Form*. Cambridge, MA; MIT Press.

Lynch, K. and G. Hack (1984) *Site Planning*. Cambridge, MA; MIT Press.

Lyons, E. (1983) Demographic correlates of landscape preference. *Environment and Behavior* 15; 487–511.

McAllister, D.M. (1982) *Evaluation in Environmental Planning*. Cambridge, MA; MIT Press.

McHarg, I. (1969) *Design with Nature*. Garden City, NY; Natural History Press.

McKechnie, G.E. (1974) *Manual for the Environmental Response Inventory*. Palo Alto, California; Consulting Psychologists Press.

McKibben, W. (1989) *The End of Nature*. New York; Random House.

Macia, A. (1975) Visual perception of landscape; pp. 279–85 in G.H. Elsner and R.C. Smardon (eds) *Our National Landscape*. Berkeley, California; Pacific Southwest Forest and Range Experimental Station.

Maier, C.S. (1992) Democracy since the French Revolution; pp. 125–54 in J. Dunn (ed) *Democracy: The Unfinished Journey*. Oxford; Oxford University Press.

Maltzahn, K.E. von (1994) *Nature as Landscape: Dwelling and Understanding*. Montreal; McGill-Queen's University Press.

Marsh, J. and J. Gardner (1978) Recreation in consumer and conserver societies. *Alternatives* 7; 25–9.

Marx, L. (1964) *The Machine in the Garden*. New York; Oxford University Press.

Maslow, A.H. (1954) *Motivation and Personality*. New York; Harper and Row.

——— (1966) *The Psychology of Science*. New York; Harper and Row.

Maslow, A.H. and N.L. Mintz (1956) Effects of esthetic surroundings I. *Journal of Psychology* 41; 247–54.

Meinig, D.W. (1971) Environmental appreciation. *Western Humanities Review* 25; 1–11.

——— (ed) (1979) *The Interpretation of Ordinary Landscapes*. New York; Oxford University Press.

——— (1982) Geography as an art. *Transactions, Institute of British Geographers* NS 8; 314–28.

Mellander, K. (1974) *Environmental Planning: Can Scale Models Help?* Berkeley, California; University of California, Institute of Urban and Regional Development Reprint 125.

Merrill, D. and J. Board (1980) Perception and recall of aesthetic quality in a familiar environment. *Psychological Research* 42; 375–90.

Middleton, M. (1987) *Man Made the Town*. New York; St. Martin's Press.

Miller, B. (1979) Seattle's Freeway Park. *American Forests* 85; 28–31.

Mintz, N.L. (1956) Effects of esthetic surroundings II. *Journal of Psychology* 41; 459–66.

Mitroff, I.I. and R.H. Kilmann (1978) *Methodological Approaches to Social Science*. San Francisco; Jossey-Bass.

Moffit, L.C. (1975) Value implications for public planning: some thoughts and questions. *Journal of the American Institute of Planners* 41; 397–405.

Moggridge, H. (1984) The working method by which the original composition of planting around L. Brown's lakes was defined. *Landscape Research* 9; 15–23.

——— (1986) The delights and problems of practice; pp. 102–13 in E.C. Penning-Rowsell and D. Lowenthal (eds) *Landscape Meanings and Values*. London; Allen and Unwin.

Moholy-Nagy, S. (1968) *Matrix of Man*. London; Pall Mall Press.

Moore, G.T. (ed) (1970) *Emerging Methods in Environmental Design and Planning*. Cambridge, MA; MIT Press.

Moore, R. (1980) Collaborating with young people to assess their landscape values. *Ekistics* 281; 128–35.

Morgan, M. (1978) Perspectives on landscape aesthetics. *Progress in Human Geography* 2; 527–32.

Moss, M.R. and W.G. Nickling (1989) Environmental and policy requirements: some Canadian examples and the need for environmental process assessment; pp. 177–94 in P. Dearden and B. Sadler (eds) *Landscape Evaluation: Approaches and Applications*. Victoria, BC; Western Geographical Series.

Motion, A. (1993) *Philip Larkin*. London; Faber.

Mumford, L. (1938) *The Culture of Cities*. London; Secker and Warburg.

——— (1964) *The Highway and the City*. London; Secker and Warburg.

——— (1966) *The Urban Prospect*. New York; Harcourt, Brace and World.

Nabhan, G.P. and S. St. Antoine (1993) The loss of floral and faunal story: the extinction of experience; pp. 229–50 in S.R. Kellert and E.O. Wilson (eds) *The Biophilia Hypothesis*. Washington, DC; Island Press.

Nairn, I. (1965) *The American Landscape*. New York; Random House.

Nasar, J.L. (1984) Visual preference in urban street scenes; a cross-cultural comparison between Japan and the United States. *Journal of Cross-Cultural Psychology* 15; 79–83.

——— (ed) (1988) *Environmental Aesthetics: Theory, Research and Applications*. Cambridge; Cambridge University Press.

Nasar, J.L., D. Julian, S. Buchman, D. Humphreys and M. Mrohaly (1983) The emotional quality of scenes and observation points: a look at prospect and refuge. *Landscape Planning* 10; 355–61.

Nash, R. (1973) *Wilderness and the American Mind*. New Haven, Conn.; Yale University Press.

Nead, L. (1992) *The Female Nude; Art, Obscenity and Sexuality*. London; Routledge.

Nelson, R. (1993) Searching for the lost arrow: physical and spiritual ecology in the hunter's world; pp. 201–28 in S.R. Kellert and E.O. Wilson (eds) *The Biophilia Hypothesis.* Washington, DC; Island Press.

Newcomb, R.M. (1979) *Planning The Past.* London; Dawson.

Newton, E. (1950) *The Meaning of Beauty.* New York; Whittlesey House.

Nicholson, M.H. (1959) *Mountain Gloom and Mountain Glory: The Development of the Aesthetics of the Infinite.* Ithaca, NY; Cornell University Press.

Nieman, T.J. (1980) The visual environment of the New York coastal zone: user preferences and perceptions. *Coastal Zone Management Journal* 8; 45–62.

Nohl, W. (1980) Open space in cities: in search of a new aesthetic. *Landscape* 28; 35–40.

Norberg-Schulz, C. (1981) *Genius Loci: Paysage, Ambiance, Architecture.* Brussels; Mardaga.

Novak, B. (1980) *Nature and Culture.* New York; Oxford University Press.

Novak, M. (1985) *Successful Aging.* Markham, Ontario; Penguin.

Olwig, K.R. (1993) Sexual cosmology; pp. 307–43 in B. Bender (ed) *Landscape: Politics and Perspective.* Oxford; Berg.

Orians, G. (1980) Habitat selection: general theory and applications to human behaviour; pp. 86–94 in J.S. Lockard (ed) *The Evolution of Human Social Behaviour.* New York; Elsevier.

—— (1986) An ecological and evolutionary approach to landscape aesthetics; pp. 3–25 in E.C. Penning-Rowsell and D. Lowenthal (eds) *Landscape Meanings and Values.* London; Allen and Unwin.

Orians, G. and L. Heerwagen (1992) Evolved responses to landscapes; pp. 111–23 in J. Barkow, L. Cosmides and J. Tooby (eds) *The Adapted Mind: Evolutionary Psychology and the Generation of Culture.* Oxford; Oxford University Press.

O'Riordan, T. (1976) *Environmentalism.* London; Pion.

Orland, B. (1988) Aesthetic preferences for rural landscapes: some resident and visitor differences; pp. 364–78 in J.L. Nasar (ed) *Environmental Aesthetics.* Cambridge; Cambridge University Press.

Parsons, R. (1991) The potential influences of environmental perception on human health. *Journal of Environmental Psychology* 11; 1–23.

Patmore, J.A. (1971) *Land and Leisure.* Harmondsworth, UK; Penguin.

Pawley, M. (1974) *The Private Future.* London; Thames and Hudson.

Pearce, S. and N. Waters (1983) Quantitative methods for investigating the variables that underlie preference for landscape scenes. *Canadian Geographer* 27; 328–44.

Pearlman, K.T. (1988) Aesthetic regulation and the courts, pp. 476–92 in J.L. Nasar (ed) *Environmental Aesthetics.* Cambridge; Cambridge University Press.

Pellegrini, C. (1991) *Evaluation Esthétique du Paysage.* Neuchatel, Switzerland; Cahiers de L'Institut de Geographie 21.

Pendse, D. and J.B. Wyckoff (1974) Environmental goods: determination of preferences and trade-off values. *Journal of Leisure Research* 6; 64–76.

Penning-Rowsell, E.C. (1973) *Alternative Approaches to Landscape Appraisal and Evaluation.* Enfield, UK; Middlesex Polytechnic Planning Research Group Report 11.

—— (1974) Landscape evaluation for development plans. *The Planner: Journal of the Royal Town Planning Institute* 60; 930–4.

—— (1975) *Alternative Approaches to Landscape Appraisal and Evaluation: Supplement.* Enfield, UK; Middlesex Polytechnic Planning Research Group Report 11, supplement.

—— (1981a) Fluctuating fortunes in gauging landscape value. *Progress in Human Geography* 5; 25–41.

—— (1981b) Assessing the validity of landscape evaluations. *Landscape Research* 6; 22–4.

—— (1989) Landscape research and practice: recent developments in the United Kingdom; pp. 243–61 in P. Dearden and B. Sadler (eds) *Landscape Evaluation: Approaches and Applications.* Victoria, BC; Western Geographical Series.

Penning-Rowsell, E.C. and D.I. Hardy (1973) Landscape evaluation and planning policy: a comparative survey in the Wye Valley Area of Outstanding Natural Beauty. *Regional Studies* 7; 153–60.

Penning-Rowsell, E.C. and D. Lowenthal (eds.) (1986) *Landscape Meanings and Values.* London; Allen and Unwin.

Penning-Rowsell, E.C. and G.H. Searle (1977) The 'Manchester' landscape evaluation method: a critical appraisal. *Landscape Research* 2; 6–11.

Penning-Rowsell, E.C., G.H. Gullett, G.H. Searle, and S.A. Witham (1977) *Public Evaluation of Landscape Quality.* Enfield, UK; Middlesex Polytechnic Planning Research Group Report 13.

Pennyfather, K. (1975) *Guide to Countryside Interpretation 2*. Edinburgh; HMSO.

Percival, A. (1979) *Understanding Our Surroundings: A Manual of Urban Interpretation*. London; Civic Trust.

Petersen, D. (ed) (1994) *Confessions of a Barbarian: Selections from the Journal of Edward Abbey*. Boston; Little, Brown.

Peterson, G.L. (1967) A model of preference: qualitative analysis of the perception of visual appearance of residential neighborhoods. *Journal of Regional Science* 7; 19–31.

———— (1975) Recreational preferences of urban teenagers; pp. 113–21 in *Children, Nature and the Urban Environment*, Proceedings of a Symposium at George Washington University, Washington DC.

Peterson, G.L. and E.S. Neumann (1969) Modelling and predicting human response to the visual recreation environment. *Journal of Leisure Research* 1; 219–37.

Phillips, S.D. and T.M. Semple (1978) Environmental dispositions and the education of environmental specialists; pp. 480–8 in: W. Rogers and W. Ittelson (eds) *New Directions in Environmental Design Research*. Washington, DC, EDRA.

Pocock, D.C.D. (ed) (1981) *Humanistic Geography and Literature*. London; Croom Helm.

Porteous, J.D. (1977) *Environment & Behavior: Planning and Everyday Urban Life*. Reading, MA; Addison-Wesley.

———— (1982) Approaches to environmental aesthetics. *Journal of Environmental Psychology* 2; 53–60.

———— (1984) Putting Descartes before dehors. *Transactions, Institute of British Geographers* NS 9; 372–3.

———— (1985) Literature and humanistic geography. *Area* 17; 117–22.

———— (1986) Intimate sensing. *Area* 18; 250–1.

———— (1988) *Degrees of Freedom*. Saturna Island, BC; Saturnalia Press.

———— (1989) *Planned to Death*. Manchester; Manchester University Press.

———— (1990) *Landscapes of the Mind*. Toronto; University of Toronto Press.

———— (1991) Transcendental experience in wilderness sacred space. *National Geographical Journal of India* 37; 99–107.

———— (1992) A loving nature: Malcolm Lowry in British Columbia; pp. 258–69 in P. Simpson-Housley and G. Norcliffe (eds). *A Few Acres of Snow: Literary and Artistic Images of Canada*. Toronto; Dundurn Press.

———— (1993) Resurrecting environmental religion. *National Geographical Journal of India* 39; 179–87.

Porteous, J.D. and J. Mastin (1985) Soundscape. *Journal of Architectural and Planning Research* 2; 169–86.

Postman, N. (1982) *The Disappearance of Childhood*. New York; Delacorte Press.

———— (1993) *Technopoly: The Surrender of Culture to Technology*. New York; Random House.

Prall, D. (1929) *Aesthetic Judgement*. New York; Crowell.

Preece, R.A. (1991) *Designs on the Landscape*. London; Belhaven.

Preiser, W.F.E. (ed) (1973) *Environmental Design Research*. Stroudsburg, PA; Dowden, Hutchinson and Ross.

Preiser, W.F.E. and K.P. Rohane (1988) A survey of aesthetic controls in English-speaking countries; pp. 422–33 in J.L. Nasar (ed) *Environmental Aesthetics*. Cambridge; Cambridge University Press.

Prentice, R. (1991) Measuring the educational effectiveness of on-site interpretation designed for tourists. *Area* 23; 297–308.

Price, C. (1976) Subjectivity and objectivity in landscape evaluation. *Environment and Planning A* 8; 829–38.

Priestley, J.B. (1955) *Journey Down a Rainbow*. London; Penguin.

Priestly, T. (1983) The field of visual analysis and resource management: a bibliographic analysis and perspective. *Landscape Journal* 2; 52–9.

Prince, H. (1981) Revival, restoration, preservation; pp. 33–49 in D. Lowenthal and M. Binney (eds) *Our Past Before Us: Why Do We Save It?* London; Temple Smith.

Punter, J.V. (1982) Landscape aesthetics: a synthesis and critique; pp. 100–23 in J.R. Gold and J. Burgess (eds) *Valued Environments*. London; Allen and Unwin.

Randall, A., B. Ives and C. Eastman (1974) Bidding games for valuation of aesthetic environmental improvements. *Journal of Environmental Economics and Management* 1; 132–49.

Rapoport, A. (1977) *Human Aspects of Urban Form*. Oxford; Pergamon.

Rapoport, A. and R. Hawkes (1970) The perception of urban complexity. *Journal of the American Institute of Planners* 36; 106–11.

Rapoport, A. and R.E. Kantor (1967) Complexity and ambiguity in environmental design. *Journal of the American Institute of Planners* 33; 210–22.

Rasmussen, S.E. (1951) *Towns and Buildings.* Cambridge, MA; MIT Press.

—— (1959) *Experiencing Architecture.* Cambridge, MA; MIT Press.

Rees, R. (1973) Geography and landscape painting. *Scottish Geographical Magazine* 89; 147–57.

—— (1975) The scenery cult; changing landscape tastes over three centuries. *Landscape* 19; 39–48.

Relph, E. (1976) *Place and Placelessness.* London; Pion.

—— (1979) To see with the soul of the eye. *Landscape* 23; 28–34.

—— (1981) *Rational Landscapes and Humanistic Geography.* London; Croom Helm.

—— (1982) The landscapes of a conserver society; pp. 47–65 in B. Sadler and A. Carlson (eds) *Environmental Aesthetics: Essays in Interpretation.* Victoria, BC; Western Geographical Series.

—— (1987) *The Modern Urban Landscape.* Baltimore; Johns Hopkins University Press.

—— (1993) Critical reflections on environmental ethics and sense of place. *National Geographical Journal of India* 39; 85–9.

Ridout, M. (1988) Scenic-beauty issues in public policy making; pp. 434–48 In J.L. Nasar (ed) *Environmental Aesthetics.* Cambridge; Cambridge University Press.

Riley, R.B. (1987) Vernacular landscapes; pp. 129–58 in E.H. Zube and G.T. Moore (eds) *Advances in Environment, Behavior, and Design 1.* New York: Plenum Press.

Rimbert, S. (1973) *Les Paysages Urbains.* Paris; Armand Colin.

Rivlin, R. and Gravelle, K. (1984) *Deciphering the Senses.* New York; Simon and Schuster.

Robinson, D.G., J.F. Wager, I.C. Laurie and A.L. Traill (1976) *Landscape Evaluation.* Manchester; University of Manchester, Centre for Urban and Regional Research.

Rock, I. and C.S. Harris (1967) Vision and touch. *Scientific American* 216; 96–104.

Rodaway, P. (1994) *Sensuous Geographies.* London; Routledge.

Roepke, H.G. (1977) Applied geography: should we, must we, can we? *Geographical Review* 67; 481–2.

Rose, M.C. (1976) Nature as an aesthetic object: an essay in meta-aesthetics. *British Journal of Aesthetics* 16; 3–12.

Rostow, W.W. (1960) *The Stages of Economic Growth.* Cambridge; Cambridge University Press.

Rowe, P. (1987) *Design Thinking.* Cambridge, MA; MIT Press.

Russell, J.A. and J. Snodgrass (1982) Emotion and environment; pp. 245–80 in D. Stokols and I. Altman (eds) *Handbook of Environmental Psychology.* New York; Wiley.

Sadler, B. and A.A. Carlson (eds) (1982) *Environmental Aesthetics: Essays in Interpretation.* Victoria, BC; Western Geographical Series.

Sadler, B. and P. Dearden (1989) Where to now? Some future directions for geographical research in landscape evaluation; pp. 265–9 in P. Dearden and B. Sadler (eds) *Landscape Evaluation: Approaches and Applications.* Victoria, BC; Western Geographical Series.

Samuels, M. (1979) The biography of landscape; pp. 51–88 in D.W. Meinig (ed) *The Interpretation of Ordinary Landscapes.* New York; Oxford University Press.

Sancar, F.H. (1985) Toward theory generation in landscape aesthetics. *Landscape Journal* 4; 116–24.

—— (1989) A critique of criticism: can the avant-garde embrace the entire landscape? *Avant Garde* 2; 78–91.

Sanguin, A-L. (1981) La géographie humaniste ou l'approche phénoménologique des lieux, des paysages et des espaces. *Annals de Géographie* 501; 560–87.

Sansot, P. (1983) *Variations Paysagères: Invitation au Paysage.* Paris; Klincksieck.

Santayana, G. (1896) *The Sense of Beauty.* New York; Collier.

Sautter, G. (1979) Le paysage comme connivence. *Hérodote* 16; 40–67.

Saw, R. and H. Osborne (1968) Aesthetics as a branch of philosophy; pp. 15–32 in H. Osborne (ed) *Aesthetics in the Modern World.* London; Thames and Hudson.

Schachtel, E.G. (1959) *Metamorphosis.* New York; Basic Books.

Schafer, R.M. (1977) *The Tuning of the World.* Toronto; McClelland and Stewart.

Schaffer, K. (1988) *Women and the Bush: Forces of Desire in the Australian Cultural Tradition.* Cambridge; Cambridge University Press.

Schama, S. (1995) *Landscape and Memory.* New York; Random House.

Scheer, E. (ed) (1979) *Paysage Sonore Urbain.* Paris; Plan-Construction.

Schumacher, E.F. (1973) *Small is Beautiful*. New York; Harper and Row.

Seamon, D. (1976) Extending the man–environment relationship: Wordsworth and Goethe's experience of the natural world. *Monadnock* 50; 18–41.

Searles, H.F. (1961) The role of the non-human environment. *Landscape* 11; 31–4.

Seaton, R. and J. Collins (1972) Validity and reliability of ratings of simulated buildings; pp. 6.10.1–6.10.12 in W.J. Mitchell (ed) *Environmental Design: Research and Practice*. Los Angeles; UCLA Press.

Sewell, J. (1993) *The Shape of the City: Toronto Struggles with Modern Planning*. Toronto; University of Toronto Press.

Shafer, E.L. (1969) Perception of natural environments. *Environment and Behavior* 1; 71–82.

Shafer, E.L. and J. Mietz (1969) Aesthetic and emotional experiences rate high with North-east wilderness hikers. *Environment and Behavior* 1; 187–97.

Shafer, E.L. and J. Mietz (1970) *It Seems Possible to Quantify Scenic Beauty in Photographs*. Upper Darby, PA; USDA Forest Service Research Paper NE–162.

Shafer, E.L., J.F. Hamilton and E. Schmidt (1969) Natural landscape preference: a predictive model. *Journal of Leisure Research* 1; 1–19.

Sharpe, G.W. (ed) (1976) *Interpreting the Environment*. New York; Wiley.

Shelden, M. (1991) *Orwell*. New York; Harper Collins.

Shepard, P. (1967) *Man in the Landscape: A Historic View of the Aesthetics of Nature*. New York; Knopf.

Shuttleworth, S. (1980) The use of photographs as an environmental presentation medium in landscape studies. *Journal of Environmental Management* 11; 61–76.

Silk, J. (1984) Beyond geography and literature. *Environment and Planning D: Society and Space* 2; 151–78.

Simms, A. (1982) Review of R.M. Newcomb (1979) *Planning the Past*. *Economic Geography* 58; 292–3.

Singer, G.L (1979) Citizens defend the urban coast. *Bulletin of the Atomic Scientists* 35; 47–52.

Smardon, R.C. and J.P. Felleman (1982) The quiet revolution in visual resource management: a view from the coast. *Coastal Zone Management Journal* 9; 211–24.

Smith, B. and H.B. Reynolds (eds) (1987) *The City as a Sacred Center*. Leiden; Brill.

Smith, P.F. (1974) *Dynamics of Urbanism*. London; Hutchinson.

—— (1977) *The Syntax of Cities*. London; Hutchinson.

Sonnenfeld, J. (1966) Variable values in space and landscape: an enquiry into the nature of environmental necessity. *Journal of Social Issues* 4; 71–82.

—— (1967) Environmental perception and adaptation level in the Arctic; pp. 42–59 in D. Lowenthal (ed) *Environmental Perception and Behavior*. Chicago; Department of Geography, University of Chicago.

Sorte, G.J. (1971) *Perception av Landskap*. Unpublished Licentiate's dissertation, Universitetsforlaget, As, Norway.

—— (1975) Methods for presenting planned environment. *Man–Environment Systems* 5; 148–54.

Southworth, M. (1969) The sonic environment of cities. *Environment and Behavior* 1; 49–70.

Spooner, D. (1992) Larkin's *Here*. *Geography* 77: 134–42.

Steinitz, C. (1970) Landscape resource analysis: the state of the art. *Landscape Architecture* 60; 101–5.

Steinitz, C. and D. Way (1969) A model for evaluating the visual consequences of urbanization; pp. 69–83 in C. Steinitz and P. Rogers (eds) *Qualitative Values in Environmental Planning*. Washington DC; Department of the Army.

Stilgoe, J.R. (1982) *Common Landscape of America, 1580 to 1845*. New Haven; Yale University Press.

Stokols, D. (1978) Environmental psychology. *Annual Review of Psychology* 29; 253–95.

Stroud, D. (1954) Eighteenth century landscape gardening; pp. 35–46 in W.A. Singleton (ed) *Studies in Architectural History*. London; St. Anthony's Press.

Symanski, R. (1994) Contested realities: feral horses in Outback Australia. *Annals of the Association of American Geographers* 84; 251–69.

Tainter, J.A. and R.H. Hamre (1988) *Tools to Manage the Past*. Fort Collins, Colorado; USDA Forest Service General Technical Report RM–164.

Thayer, R.L. (1976) Visual ecology: revitalizing the aesthetics of landscape architecture. *Landscape* 20; 37–43.

Thayer, R.L., R.W. Hodgson, L.D. Gustke, B.G. Atwood and J. Holmes (1976) Validation of a natural landscape preference model as a predictor of perceived landscape beauty in photographs. *Journal of Leisure Research* 8; 292–9.

Thompson, P. (1978) *The Voice of the Past.* Oxford; Oxford University Press.

Thornes, J.E. (1979) Landscape and clouds. *Geographical Magazine* April; 492–501.

Tiffany, S. and K. Adams (1985) *The Wild Woman; An Inquiry into the Anthropology of an Idea.* Cambridge, Mass.; Schenkman.

Tiger, L. (1992) *The Pursuit of Pleasure.* Boston; Little, Brown.

Tinker, J. (1974) The end of the English landscape. *New Scientist* 5 December, 722–7.

Tippet, M. and D. Cole (1977) *From Desolation to Splendour: Changing Perceptions of the British Columbia Landscape.* Toronto; Clarke Irwin.

Tooby, J. and L. Cosmides (1990) The past explains the present: emotional adaptations and the structure of ancestral environments. *Ethology and Sociobiology* 11; 375–424.

Traill, A.L. (n.d.) *Statistical Appraisal of Existing Techniques.* Manchester; University of Manchester Landscape Evaluation Research Project Occasional Paper 39.

——— (1978) The Manchester landscape evaluation method: a reply to E.C. Penning-Rowsell and G.H. Searle. *Landscape Research* 3; 2–4.

Truax, B. (1984) *Acoustic Communication.* Norwood, NJ; Ablex.

Tuan, Y-F. (1961) Topophilia: or, sudden encounter with the landscape. *Landscape* 11; 29–32.

——— (1974) *Topophilia.* Englewood Cliffs, NJ; Prentice Hall.

——— (1977) *Space and Place.* Minneapolis; University of Minnesota Press.

——— (1978) Literature and geography; pp. 194–206 in D. Ley and M. Samuels (eds) *Humanistic Geography.* London; Croom Helm.

——— (1982) *Segmented Worlds and Self.* Minneapolis; University of Minnesota Press.

——— (1984) *Dominance and Affection: The Making of Pets.* New Haven; Yale University Press.

——— (1989) Surface phenomena and aesthetic experience. *Annals of the Association of American Geographers* 79; 233–41.

——— (1993) *Passing Strange and Wonderful: Aesthetics, Nature, and Culture.* Washington, DC; Island Press.

——— (1994) personal communication.

Tunnard, C. and B. Pushkarev (1963) *Man-Made America: Chaos or Control?* New Haven, Conn.; Yale University Press.

Turner, J.R. (1975) Application of landscape evaluation: a planners' view. *Transactions, Institute of British Geographers* 66; 156–61.

Turner, V.W. (1988) *Forest of Symbols.* Ithaca, NY; Cornell University Press.

Udall, S. (1963) *The Quiet Crisis.* New York; Basic Books.

Ulrich, R.S. (1974) *Scenery and the Shopping Trip.* Ann Arbor, Michigan; University of Michigan Geographical Publications.

——— (1976) Urbanization and garden aesthetics. *Longwood Seminars* 8; 4–8.

——— (1979a) Visual landscape preference: a model and application. *Man–Environment Systems* 7; 279–99.

——— (1979b) Visual landscape and psychological well-being. *Landscape Research* 4; 17–19.

——— (1981) Natural versus urban scenes: some psychophysiological effects. *Environment and Behavior* 13; 523–56.

——— (1983) Aesthetic and affective response to natural environment; pp. 88–125 in I. Altman and J.F. Wohlwill (eds) *Behavior and the Natural Environment.* New York; Plenum.

——— (1984) View through a window influences recovery from surgery. *Science* 224; 420–1.

——— (1993) Biophilia, biophobia, and natural landscapes; pp. 73–137 in S.R. Kellert and E.O. Wilson (eds) *The Biophilia Hypothesis.* Washington DC; Island Press.

Ulrich, R.S. and D. Addoms (1981) Psychological and recreational benefits of a neighborhood park. *Journal of Leisure Research* 13; 43–65.

Uzzell, D. (ed) (1989) *Heritage Interpretation.* London; Belhaven.

Valentine, C.W. (1962) *The Experimental Psychology of Beauty.* London; Methuen.

Veitch, J.A. (1990) Office noise and illumination effects on reading comprehension. *Journal of Environmental Psychology* 10; 209–17.

Veitch, J.A., R. Gifford and D.W. Hine (1991) Demand characteristics and full spectrum lighting effects on performance and mood. *Journal of Environmental Psychology* 11; 87–95.

Venturi, R. (1966) *Complexity and Contradiction in Architecture.* New York; Museum of Modern Art.

Venturi, R., D.S. Brown and R. Izenhour (1972) *Learning from Las Vegas.* Cambridge, MA; MIT Press.

Wachs, M. (ed) (1985) *Ethics in Planning.* New Brunswick, NJ; Rutgers University Press.

Walker, R. (1981) A theory of suburbanization; pp. 383–430 in M. Dear and A. Scott (eds) *Urbanization and Urban Planning in Capitalist Societies*. New York; Methuen.
—— (1988) Review of E. Relph (1987) *The Modern Urban Landscape. Canadian Geographer* 32; 368–70.
Ward, C. and A. Fyson (1973) *Streetwork: The Exploring School*. London; Routledge and Kegan Paul.
Watson, W. (1983) The soul of geography. *Transactions, Institute of British Geographers* NS 8; 385–99.
Weddle, A.E. and J. Pickard (1969) Techniques in landscape planning. *Journal of the Royal Town Planning Institute* 55; 387–9.
Weinstein, N.D. (1976) The statistical prediction of environmental preferences: problems of validity and application. *Environment and Behavior* 8; 611–26.
Weiss, M. (1987) *The Rise of the Community Builders*. New York; Columbia University Press.
Westland, G. (1967) The psychologist's search for scientific objectivity in aesthetics. *British Journal of Aesthetics* 7; 350–7.
Wheatley, P. (1971) *The Pivot of the Four Quarters*. Chicago; Aldine.
Wheeler, K. and B. Waites (eds.) (1976) *Environmental Geography*. London; Hart-Davis Educational.
Whitbread, M. (1977) Problems of measuring the quality of city environments; pp. 92–110 in L. Wingo and A. Evans (eds) *Public Economics and the Quality of Life*. Baltimore; Johns Hopkins University Press.
White, L. (1967) The historical roots of our ecologic crisis. *Science* 155; 1203–7.
Whitehead, C. (1989) Review of J.L. Nasar (1988) *Environmental Aesthetics. Journal of Environmental Psychology* 9; 257–62.
Whittick, A. (1970) Aesthetics of urban design. *Journal of the Town Planning Institute* 58; 332–7.
Wildavsky, A. (1967) Aesthetic power or the triumph of the sensitive minority over the vulgar mass. *Daedalus* 96; 1115–28.
Williams, R. (1973) *The Country and the City*. New York; Oxford University Press.
Williams, S.H. (1954) Urban aesthetics. *Town Planning Review* 25; 95–113.
Wilmot, D. (1990) Maslow and Mintz revisited. *Journal of Environmental Psychology* 10; 293–312.
Wilson, A. (1991) *The Culture of Nature*. Toronto; Between the Lines.
Wilson-Hodges, C. (1978) *The Measurement of Landscape Aesthetics*. Toronto; University of Toronto, Department of Geography Environmental Perception Working Paper 2.
Winkel, G. (1973) Visual attributes of the environment: introduction; pp. 60–1 in W.F.E. Preiser (ed) *Environmental Design Research*. Los Angeles; UCLA Press.
Winkel, G., R. Malek and P. Theil (1970) A study of human responses to selected roadside environments; pp. 224–40 in H. Sanoff and S. Cohn (eds) *EDRA 1: Proceedings of the First Environmental Design Research Association Conference*. Raleigh, NC; North Carolina State University.
Winter, R. (1978) *Le Livre des Odeurs*. Paris; Seuil.
Wohlwill, J.F. (1976) Environmental aesthetics: the environment as source of affect; pp. 37–86 in I. Altman and J.F. Wohlwill (eds) *Human Behavior and Environment*. New York; Plenum.
—— (1978) What belongs where: research on the fittingness of man-made structures in natural settings. *Landscape Research* 3; 3–8.
—— (1982) The visual impact of development in coastal zone areas. *Coastal Zone Management Journal* 9; 225–48.
Wohlwill, J.F. and I. Kohn (1976) Dimensionalizing the environmental manifold; pp. 19–54 in S. Wapner, S.B. Cohen and B. Kaplan (eds) *Experiencing the Environment*. New York; Plenum.
Wolfe, T. (1981) *From Bauhaus to Our House*. New York; Farrar, Strauss, Giroux.
Wood, C.J.B. (1989) Colour and landscape; pp. 101–17 in P. Dearden and B. Sadler (eds) *Landscape Evaluation*. Victoria, BC; Western Geographical Series.
Wood, D. and R. Beck (1990) Tour personality. *Journal of Environmental Psychology* 10; 177–208.
Woodcock, D.M. (1982) *A Functionalist Approach to Environmental Preference*. Unpublished PhD thesis, University of Michigan at Ann Arbor.
—— (1984) A functionalist approach to landscape preference. *Landscape Research* 9; 24–7.
Wordsworth, W. (1977) *The Poems*, vol I (ed. J.O. Hayden). Harmondsworth; Penguin, pp. 356–7.

Wright, G. (1974) Appraisal of visual landscape qualities in a region selected for accelerated growth. *Landscape Planning* 1; 307–27.

Wright, J.B. and S.G. Hilts (1993) An overview of voluntary approaches to landscape conservation in the United States and Canada. *Operational Geographer* 11; 10–14.

Yeomans, W.C. (1983) *Visual Resource Assessment: A User Guide.* Victoria, BC; Ministry of Environment Manual 2.

Younghusband, F. (1920) Natural beauty and geographic science. *Geographical Journal* 56; 1–13.

Zeisel, J. (1981) *Inquiry by Design.* Monterey, Calif.; Brooks-Cole.

Ziegler, A. (1971) *Historic Preservation in Inner-city Areas: A Manual of Practice.* Pittsburgh; Van Trump, Ziegler and Shane.

Zube, E.H. (1973) Scenery as a natural resource: implications of public policy and problems of definition, description and evaluation. *Landscape Architecture* 63; 126–32.

——— (1974) Cross-disciplinary and intermode agreement on the description and evaluation of landscape resources. *Environment and Behavior* 6; 69–89.

——— (1976a) Perception of landscape and land use; pp. 87–122 in I. Altman and J.F. Wohlwill (eds) *Human Behavior and Environment.* New York; Plenum.

——— (1976b) Landscape aesthetics: policy and planning in the United States; pp. 68–79 in J. Appleton (ed) *The Aesthetics of Landscape.* Didcot, UK; Rural Planning Services Publication 7.

——— (ed) (1980) *Social Sciences, Interdisciplinary Research, and the US Man and the Biosphere Program.* Washington DC; State Department.

——— (1991) Environmental psychology, global issues, and local landscape research. *Journal of Environmental Psychology* 11; 321–34.

Zube, E.H. and D.G. Pitt (1981) Cross-cultural perceptions of scenic and heritage landscapes. *Landscape Planning* 8; 69–87.

Zube, E.H., T. Anderson and D.H. Pitt (1973) Measuring the landscape: perceptual responses and physical dimensions. *Landscape Research News* 1; 4–6.

Zube, E.H., R.O. Brush and J.G. Fabos (eds) (1975) *Landscape Assessment: Values, Perceptions, and Resources.* Stroudsburg, PA; Dowden, Hutchinson and Ross.

Zube, E.H., D.G. Pitt and T.W. Anderson (1974) *Perception and Measurement of Scenic Resources in the Southern Connecticut River Valley.* Amherst, MA: University of Massachusetts Institute for Man and the Environment Publication R–74–1.

Zube, E.H., D.G. Pitt and T.W. Anderson (1975) Perception and prediction of scenic resource values of the North-east; pp. 151–67 in E.H. Zube, R.O. Brush and J.G. Fabos (eds) *Landscape Assessment.* Stroudsburg, PA; Dowden, Hutchinson and Ross.

Zube, E.H., D.G. Pitt and G.W. Evans (1983) A life-span developmental study of landscape assessment. *Journal of Environmental Psychology* 3; 115–28.

Zube, E.H., J.L. Sell and G. Taylor (1982) Landscape perception: research, application and theory. *Landscape Planning* 9; 1–33.

Zube, E.H., J. Vining, C.S. Law and R.B. Bechtel (1985) Perceived urban residential quality: a cross-cultural bimodal study. *Environment and Behavior* 17; 327–50.

INDEX

Abbey, E. 78, 167
Ackerman, D. 5, 41, 147
activism, activists 3, 129, 151–87, 247;
    citizen 162–8; design 154–61; direct
    164–7; ecotage 167–8; humanist
    251–3; landscape 154–61; literary
    152–4: profiles of 165–6; protests of
    164–6; radical 254–9
adaptation-level theory 125
Addison, J. 19–20, 63, 76
advertising 5, 258
aesthetic: activism 21–2; beauty 19, 24;
    critique 249–59; justice 22; preference
    21; welfare 21–2; wealth 21–2
aesthetics 19–21; of body 5–6; concepts
    21–2; of eighteenth century 19–20;
    experimental 24; formal 22; history of
    19–23; of houses 6; medieval 19;
    modern 19; neglect of 20; origins
    24–30; philosophy of 20–3; sensory
    22; symbolic 22
age 126–7
Aiken, S.R. 247, 249
Alexander, C. 215, 247
Alps 63–4, 75–6
Anderson, L.M. 232
appearance, personal 5–6
Appleton, J. 25–30, 120, 137, 208, 250,
    262
applicability gap 13, 15, 18, 262
architects, architecture 28–9, 81, 92–100,
    127, 156, 247; criticism 251–2;
    modernist 96–8, 128, 215–30;
    post-modernist 99–100, 128, 215–30,
    237–9
Aristotle 5, 19
Arnstein's ladder of participation 164

art 19–23, 26, 51–73, 121, 235–6, 251;
    British 68–72; criticism 19–22, 251;
    Dutch 59–60; eighteenth-century
    60–6; Italian 55–8; medieval 51–5;
    picturesque 66–7; public 235–6;
    Renaissance 55–60; Surreal 73–4
Assassination of Paris (Chevalier) 234
atavistic sensitivity 25
Austen, J. 63, 250
Australia 106–7, 125, 163, 235
autobiography 244–5

Bachelard, G. 245
Bacon, R. 55
Bagley, M.D. 232
Bali 125
Ballard, J.G. 257
Bauhaus 88, 96, 170
Baumgarten, G. 19
Beardsley, M.C. 21–2
beauty 19–23, 61–2, 169–70, 176, 230,
    252, 263
Beer, A.R. 233–4
Berkeley Environmental Simulator 141,
    226, 228–30
Berger, J. 247, 256
Berleant, A. 11, 215, 217, 243, 251
Berlyne-Wohlwill experimental
    approach 118–20
Berry, W. 152
Betjeman, J. 152–4
billboards 163, 169, 171–2, 177
Blake, P. 158, 172
blandscapes 154–6, 253
Blenheim Park, Oxfordshire 208–10
Bourassa, S.C. 11, 251
Brace, P. 172–3

Bradford, W. 76
brain, functions of 27–9, 31, 36
Britain 6, 60–72, 152–4, 178–92, 208–10, 220–3, 234–7, 245, 251, 256; national parks 178–9; public participation 178–82
British Columbia 9–10, 101, 107–9, 134, 154
Broughton, R. 169–70
Buber, M. 249
Buchanan, P. 257
Buckley, W.F. 164
Bunyan, J. 52
Burchard, J.E. 234
Bureau of Land Management (US) 209, 211
Burke, E. 19, 63
Burke, G. 234–5
Burnet, T. 75
Burnham, D. 103

Cadillac-Fairview Corporation 273–9
camera, invention of 73
Canada 9–10, 36, 158–9, 186–7, 212–14, 235–9, 246
Canadian Broadcasting Corporation 187
Carlson, A. 142–3, 251
Carr, E. 47, 107
Carruth, D.B. 174
Chairs for Lovers (King) 246
Chalmers, D. 10–11, 15
Chandler, H.P. 169
Charles, Prince of Wales 160, 215
Child, I.L. 114–15
childhood, children 36–7, 72, 126–7, 245–6, 249–51, 265
citizen action 162–8, 230
city, cities 88–100, 106–7, 215–40, 253; baroque 90–3; 'cityness' 215–16; eighteenth-century 92–4; medieval 54, 90–1, 93; modern 96–9, 215–40; planning 215–40; post-modern 99–100; Victorian 92, 95–6
Civic Amenities Act (UK) 178, 182
Clark, K. 73
class, social 7, 254–8
Claude glass 66, 105, 142
Clout, H. 205, 207
Coastal Zone Management Act (US) 167, 178

Cobb, R. 244
colour 32–5, 219–20
communication 3, 234, 262
complexity 119–21, 132, 147, 218, 230
'Complexity and ambiguity in environmental design' (Rapoport and Kantor) 28
Complexity and Contradiction in Architecture (Venturi) 231
composition, design 220–30
concrete 96–7
conservation 103, 105, 180–6, 236–40
Constable, J. 68–9
consumerism 5–8, 264–5
Cooper Marcus, C. 233, 244
Costa Rican National Parks Service 243
countryside, see rural landscape
Coventry–Solihull–Warwickshire Planning Group 198–9
Cowper, W. 101
Craik, K. 193, 200
critique, aesthetic 249–59
Crosby, T. 256
Crowe, S. 192
Cuba 157
Cullen, G. 218, 220–6, 247
culture 124–6
Culture of Nature, The (Wilson) 243

Darwin, C. 73
Davies, R. 16–17
Dearden, P. 124, 202, 207–8, 212
Debord, G. 253–4
De Quincey, T. 20
design: see planning; urban design
Dickie, G. 114
disneyfication 154
documentary films 186–7

Earth Day 241
Earth First! 167–8
economics 87, 162, 241
ecotage 151, 167–8
education 192, 236, 241–8, 250–1, 263–4
Eliot, T.S. 45, 156, 265
elites, elitism 151, 162–4, 253–8; see also power
environmental: autobiography 242, 244–5; criticism 251–2; detachment 249; education 192, 236, 241–8,

250–1; interpretation 242–3;
    midwives 253; odour 37; personalities
    128–31; religion 264–5; sensitivity
    training 242, 245–7
environmental aesthetics xv–xviii, 3–19;
    comparison of schemata 15–18;
    context 6–10; future of 262; need for
    structuring 11; structural schemata
    10–15; theoretical vacuum 10;
    theories of 25–30
Environmental Aesthetics (Nasar) 11,
    132, 233
Environmental Preference
    Questionnaire 128
Environmental Response Inventory 129
environmentalism 241, 247
environments of persistent appeal 252
epiphanies 8, 72, 145–6
eternalism 241, 247–9
ethics 8–10, 258, 263–5
Evans, E. 77
experimentalists 3, 113–47; approaches
    113–15; Berlyne-Wohlwill approach
    118–20; cultural variables 124–6;
    Kaplan approach 120–4; methods
    139–47; personality variables 128–31;
    social variables 126–8; theory 143–4;
    value of 147
exploration 77, 107, 118, 121, 245
extinction of experience 258
Exxon Valdez xvi, 256

Fabos, J.G. 139, 194, 208
façadism 101, 219, 237, 239
Faludy, G. 158–9
farmland: *see* rural landscape
featurism 105
Federal Highway Administration (US)
    209, 211
Fines, K.D. 196, 198, 201
flowers 88, 106
Ford, L. 236–7
forests, foresters 52, 76–9, 128, 133
Forest Service (US) 172, 209, 211
Forster, E.M. 264
Francis, M. 233

Gaia 241
Galbraith, J.K. 230
games 230, 245–6

gardeners 61, 82–6; Bridgeman 83;
    Brown 84–6, 209; Kent 61, 83–5; Le
    Notre 82; Repton 84, 106
gardens 6, 27, 54, 61, 81–9, 106–7;
    therapeutic value of 134
gender 27, 63, 258
genetics 25–8
geoautobiography 244
geography, geographers 145, 156, 191,
    194, 239, 250–1
Gilg, A.W. 198
Gilpin, Rev. W. 63–6
Goethe, J.W. von 68, 261
Goodey, B. 240, 243, 245–6
gothick 63–5
Grand Canyon 103, 195
Grand Tour 61, 63, 68, 76, 82
Green parties 151, 168
Grosjean, G. 202, 207
Gussow, A. 174–5

ha-ha 83–4
habitat theory 25–6
Haida 9
Halprin, L. 246–7
Hamill, L. 207, 257–8
Harris, T. 77
Hawthorne, N. 77, 80–1
Heath, T.F. 248
hedonic value 118–19
Helphand, K. 244–5
Hepburn, R.W. 22–3
heritage conservation 105, 192, 236–40,
    242–3
hierarchy of needs (Maslow) 7–8, 265
Highway Beautification Act (US) 177
homogenization 154
Hopkins, G.M. 3, 45, 113, 151, 191
Horace 79
horror, appreciation of 63,75–6
Hull (UK) 153, 157, 236
humanists 3,45–109, 113, 151, 249–53;
    approach 46–50; critique 249–53
Huxtable, A.L. 160

Im 232
index of compatibility 120
intangibles 8–10
intimate sensing 68
Inuit 32, 125

Israel 234, 242
Itami 209, 211
I–Thou attitude 249–50

Jakle, J. 218–19
Japan 38, 84, 125
Jefferson, T. xvi, 6, 79
Jung, C. 8, 17, 263
Jungst, P. 230, 245

Kant, I. 19, 63
Kaplan, R. and S. 26, 120–4, 126–8,
    132–4, 139, 144
Keats, J. 19, 263
Kennedy, J.F. 6
kinaesthesia 38
Kluckner, M. 237–9
Knight, R.P. 63, 89, 209
Krutch, J.W. 78
Kurtz, J. 256

landscape 26–7, 47, 50; appraisal
    methods 194–214; appreciation
    47–50, 249–51; architects 127, 219; art
    55–60; assessment 174, 193–214;
    biography 244, 254; landform 196;
    moral 254; preferences 121–31;
    preference appraisals 201–8; types
    75–100
landscape evaluation methods 202–4,
    208–14; evaluative 196–200; expert
    195–6; interview 201–2; literature of
    257–14; nonevaluative 195;
    user-preference 201–2
Landscape Research Group 209
Landscape Year 240
Larkin, P. 152–4
Las Vegas 29
law, legislation 169–78, 180–7, 241, 247;
    administrative 172; American 169–72,
    176–8; British 178–82; Canadian
    184–7; LILCO case 173–5; NEPA
    177; nuisance 171; and public policy
    176–87; US Supreme Court 170, 173;
    zoning 171–2
Learning Through Landscape
    programme 246
Leopold, A. 78, 248
Leopold, L.B. 195–6, 207–8, 257
Lewis, P.F. 49, 163, 250–1

limbic system 27–9, 36, 138
Linton, D. 196, 205
Litton, B. 195–6, 205
London 96, 98, 160–1, 235, 236
Long Island Lighting Company
    (LILCO) 173–5
Lorenzetti, A. 55–6
Lorrain, C. 55–6, 64, 66, 82
Lowenthal, D. 101, 103, 105, 159, 181,
    208, 236, 252
Lowry, M. 107, 152, 187
Lozano, E. 231
Lynch, K. 6, 116, 218–20, 226

machines 79–81, 264
McKibben, W. 152, 258
Managua, Nicaragua 232
Manchester Landscape Evaluation
    Method 198, 200
Marx, L. 79, 81
Maslow, A. 7–8, 115–16, 249, 251, 265
Meinig, D.W. 48, 249–51
middle landscape 78–81
Middleton, M. 239
Mitroff, I. 16–18
Moggridge, H. 87, 209–10
monkeywrenching 167
monumentalism 90, 96–8
Moresby Island, Queen Charlottes 9–10
Morgan, M. 29–30
Morris, W. 180
Moss, M.R. 212–14
mountains 54, 63–4, 71, 75–8
Muir, J. 78, 248
Multiple Use Sustained Yield Act (US)
    177
Mumford, L. 158
museums 242–3
music 24, 73, 251
Muzak 154
mystery 64, 84, 113, 121–3, 132, 147, 218

Nairn, I. 158
Nasar, J. 11, 132, 233
national aesthetic wealth 21–2
National Environmental Policy Act
    (NEPA, US) 177
National Geographic 71
National Historic Landmark
    Preservation Act (US) 177

national parks 78, 176, 178–9, 242–3
National Parks and Access to the
    Countryside Act (UK) 178
National Parks Service (US) 176
National Wild and Scenic River Systems
    Act (US) 176–7
nature 22–3, 47–5, 75–8, 82, 128;
    preferred to urban scenes 132–4;
    worship 68, 72, 77–8
nature tranquillity hypothesis 135–7, 147
neuropsychology 27–9
New Jersey (US) 165–7, 171–2
New Jerusalem, The (Kutcher) 234
Newton, E. 24
New York 160, 170, 173–5
New Yorkization 228–30
North America 5,12, 36, 48–50, 76–80,
    104, 236–40, 251
novels 63, 249–50

Olmstead, F.L. 78, 135
'One Hundred and One Questions for
    Reading a Landscape' 247, 250
optimization principle 119
Orians, G. 26–7, 30
Orwell, G. 153, 253
Oxford 221
Outdoor Challenge Program (US) 144
Outdoor Recreation Resources Review
    Committee (US) 176

painters: see art
Palladio, Palladianism 92
Paris 96, 160
park, parkland 81–6, 242, 254, 256
Parsons, R. 137–8
pastoralism 61–2, 72, 78–81, 101
pathetic fallacy 62
Pattern Language, A (Alexander) 247
Paving Paradise (Kluckner) 237
Peacock, T.L. 63
peak experience: see epiphany
Peake, M. 63
Penning-Rowsell, E.C. 200, 202–4,
    208–9
perceived environmental quality index
    205
personality 128–31
Petrarch 53–4, 75
Phillips, S.D. 129–30, 132

physiology 28–30, 136
picturesque 63–6
Pig Principle 170
place, placelessness 46, 154–5, 237, 252–3
planners, planning 3, 127, 191–240,
    252–3, 257; critique 156; landscape
    aesthetic 193–214; public
    participation in 164, 202, 204, 240;
    and public policy 176–87; and public
    preference 201–2, 215, 217, 230–4,
    240, 252; urban 215–40
poetry, poets 52, 78, 80, 81, 83, 101,
    152–4, 251; as activists 152–4;
    Betjeman 152–4; Greek and Roman
    78–80; Larkin 152–4; pastoral 62;
    Pope 62, 83, 103, 182; romantic 72,
    Sandburg 81; wasteland 73
Pope, A. 62, 83, 103, 182
Postman, N. 253, 257–8
post-modernism 99–100, 128, 215, 237–9
Poussin, N. 55, 57, 64
power 253–8
Prall, D. 10, 21
Preiser, W. 235–6
Prentice, R. 242–3
preservation 180–6, 236–40
Price, U. 63
Priestley, J.B. 154
proletarian revolution 254
prospect–refuge theory 25–7
psychiatry 8, 17, 249
psychology, psychologists 113, 115–16,
    118–47, 232, 265
public participation 164, 202, 204, 240
public policy 176–87; American 176–8;
    British 178–83; Canadian 184–7
public preferences 201–2, 215, 217,
    230–4, 240, 252
Punter, J.V. 11–12, 15
Puritans 76–7

quality of environment 6–7
quality of life 6–7

radical critique 253–8
Rees, R. 250
religion 51–2, 68, 75–9, 241, 247–9,
    264–5
Relph, E. 46, 154–6, 237, 247, 250, 253
Renaissance 19, 55–60

Repton, H. 84, 106
rigour-relevance schema 13–16, 18
Romanticism 23, 63, 66–73, 77–8, 104, 107, 250
Rosa, S. 52, 64
Rousseau, J.-J. 66, 77
ruins 64–5, 84
rural landscapes 78–81, 101–3, 133, 174–5, 191, 193, 240, 253
Ruskin, J. 55, 68, 75, 180, 250, 264

Samuels, M. 244
Sancar, F.H. 143–4, 251
San Francisco 227–30, 236–7
Santayana, G. 264
Saturna Island 33, 40
savannah 26, 124, 133
scenery 48, 132–8, 196–7, 259; therapeutic value of 132–8
Scenic Beauty Estimation Method 232
scenic resources of Scotland 196
Schumacher, E.F. 241
science, scientists 16–18, 250, 252, 257, 262–4; typology of 16–18
Scotland 196–8
Seamon, D. 72
Searles, H. 249
self-help books 5,8
self-revelation 244–5
Sense of Beauty, The (Santayana) 264
senses 31–41
sensitivity training 245–7
sensory walks 245–7
Sewell, J. 221
Shafer, E.L. 201
Sharpe, T. 221
Sierra Club 78, 104
Slough, 153
smell 31, 36–7, 169
Smith, P.F. 27–30, 46, 262
socialist critique 254
Society of Antiquaries 180
Society for the Preservation of Ancient Buildings 180
sound 33–6, 169
standard of living 6–7
Standards for Planning Water and Land Resources (US) 178, 193
Streep, M. 186
streets 90, 92–3, 239, 246, 248

Stroud, D. 87
sublime, sublimity 63, 73, 76–8, 103–4
Suzuki, D. 187
Switzerland 75, 182, 202
Syntax of Cities (Smith) 27–9, 46

tactility 37–8
taste(s) 20–1, 36, 45–6, 51–74, 101–9; American landscape 103–5; Australian landscape 106–7; British Columbian landscape 107–9; English landscape 101–3; history of 51–74
Technopoly 257
television 258, 264
theology : see religion
theophany, see epiphanies
Thoreau, H.D. 68, 72, 77–8, 248
Tillich, P. 249
total resource design 205
tourism, tourists 68, 71, 79, 82, 131, 142, 252–4, 258
Town and Country Planning Act (UK) 178
Town and Country Planning Association 246
townscape 88–100, 121, 215, 220–6, 253–4; modernist 96–8; post-modernist 99–100
Townscape (Cullen) 220–1
town scores 246–7
town trail 245–6
transcendental experience: see epiphany
transportation 134–5, 165, 246
trees 101, 106–7, 137, 243
Tuan, Y-F. 11, 20, 154, 250, 252
Turner, J.M.W. 68–71
Tunnard, C. 158, 256–7

Udall, S. 157
Ulrich, R.S. 26–7, 30, 123, 134–7
uniqueness ratios 195
United States 5–6, 49, 103–5, 125, 154, 209–11, 219–20, 234–7, 256
urban aesthetic design and planning 215–40; classical approaches 218–20; as composition 220–30; controls 235–40; heritage conservation 236–40; planning guides 237; for public preferences 230–4; qualitative 215–30; warfare 172–3

urban designers 215, 217–30, 253–4
urban environmental quality studies 231–3
urban scale models 226, 228

value(s) 151, 118–19, 252–3
valued environments 252–3
Vancouver, BC 97, 134, 236–7
vegetation, benefits of 135–8
veneerism 237–9
Venturi, R. 29, 100, 231
Versailles 82
Victoria, BC 99, 107, 136, 154, 227, 236–9
Victorian Society (UK) 182
views, sequencing of 281–21
vision 32–5, 249–51
visual: absorption capacities 205; education 250–1; quality 218–21, 230; quality objectives 205; resource management 202–6

Wales 71, 242
Walpole, H. 63, 82–4
water, benefits of 135–8
Waterfront Coalition of Hudson and Bergen Counties (US) 165, 167
Ways of Seeing (Berger) 247, 256
Westland, G. 114–15

Whitbread, M. 231–2
White House Conference on Natural Beauty (1965) 176
Wildavsky, A. 162–3
Wilde, O. xvii, 20, 47
wilderness 54, 76–8, 87, 103–5, 117, 131, 145–6, 247–8
Wilderness Act (US) 78, 177
Wilderness Laboratory, A 144
Williams, R. 254–5
Williams, S.H. 230, 242–3
Wilson, A. 186, 243
Wilson-Hodges, C. 205–6
Wohlwill, J.F. 118–20, 139, 141–2
Wolfe, T. 128
Wordsworth, W. 23, 68, 72, 76, 80, 146, 265
World Wildlife Fund 109
Wye Valley (UK) 202–3

Yellowstone National Park 176
York 90, 93, 103, 222–3
Yosemite Valley 176

Zoning 37, 88, 171–2
Zube, E.H. 12–13, 15, 126–7, 139, 142, 176, 193, 202, 207–9, 212, 233, 240
Zuckerman Inventory of Personal Reactions 135–6